PUBLIC ENGAGEMENT
AND EMERGING TECHNOLOGIES

PUBLIC ENGAGEMENT
AND EMERGING TECHNOLOGIES

Edited by Kieran O'Doherty and Edna Einsiedel

UBCPress · Vancouver · Toronto

21 20 19 18 17 16 15 14 13   5 4 3 2 1

Printed in Canada on FSC-certified ancient-forest-free paper that is processed chlorine- and acid-free.

___

Library and Archives Canada Cataloguing in Publication

Public engagement and emerging technologies / edited by Kieran O'Doherty and Edna Einsiedel.

Includes bibliographical references and index.
Issued also in electronic formats.
ISBN 978-0-7748-2460-6 (bound); ISBN 978-0-7748-2461-3 (pbk.)

1. Biotechnology – Technological innovations – Social aspects. 2. Biotechnology – Technological innovations – Moral and ethical aspects. 3. Biotechnology – Technological innovations – Government policy. 4. Biotechnology – Technological innovations – Political aspects. I. O'Doherty, Kieran, 1974- II. Einsiedel, Edna F.

| TP248.23.P82 2012 | 303.48'3 | C2012-906289-8 |

Canadä

UBC Press gratefully acknowledges the financial support for our publishing program of the Government of Canada (through the Canada Book Fund), the Canada Council for the Arts, and the British Columbia Arts Council.

This book has been published with the help of a grant from the Canadian Federation for the Humanities and Social Sciences, through the Awards to Scholarly Publications Program, using funds provided by the Social Sciences and Humanities Research Council of Canada.

The authors acknowledge the support from Genome Canada and the Valgen project, which made possible the symposium on publics and emerging technologies from which this book was developed.

UBC Press
The University of British Columbia
2029 West Mall
Vancouver, BC V6T 1Z2
www.ubcpress.ca

Contents

PUBLIC ENGAGEMENT
AND EMERGING TECHNOLOGIES

BKH. Introduction

KIERAN O'DOHERTY AND EDNA EINSIEDEL

The development of new technologies has been increasing at a rate that is difficult to fathom. Not only do these new technologies and the science that they are based on have their own inherent complexities, but they also raise novel and often unanticipated challenges relating to their integration in society. Examples abound of technologies characterized by controversy and ethical challenges. In particular, in the area of biotechnology, examples such as stem cells, genetically modified food, and biobanks, to name but a few, illustrate the complexity of the social context that arises in response to these emerging technologies.

Over the past few decades, recognition of these challenges has led to different ways of involving publics in the assessment and regulation of such emerging technologies. In some instances mandated by regulatory bodies, such engagement is motivated by questions of benefit sharing, assessing the impacts of new technologies on various aspects of society, and attempting to reconcile pluralistic value sets in regulatory frameworks. How this has been accomplished, who precisely has been consulted and when, and in which aspects of technology development publics have been involved have not only varied geographically but also evolved over time. Factors such as existing regulatory frameworks specific to jurisdictions, local cultural contexts, and the nature of particular technologies have ensured that there is no single pathway that characterizes the relationship between publics and new technologies.

Although a growing corpus of academic work on the topic has accumulated over the past few decades, there is a gap in theorizing many of the aspects of engaging publics in emerging technologies. This gap is not surprising because over time the subject has been approached from several different disciplines and combinations of disciplines, including political science, sociology, applied ethics, communication theory, and psychology. Consequently, not only methods but also disciplinary assumptions and epistemological foundations have varied.

In this book, we seek to put the subject of publics and their engagement in emerging technologies on a stronger theoretical footing. With a primary, though not exclusive, focus on genomic technologies, we draw on the expertise of leading theorists and practitioners in the field to provide overviews of methods and approaches and normative accounts of the processes underlying public engagement. Although most of our contributors blend both empirical and conceptual components in their own work, we focus here on the conceptual and use the empirical mainly to support the development of sound theoretical accounts. That is, we want to avoid adding yet more case studies of public engagement exercises to the literature; our aim is to contribute a novel and useful synthesis of well-supported theory pertaining to the different aspects of public engagement in emerging technologies.

The multi-disciplinary scholarship represented in this volume lies at the intersection of two fields: public participation and the social and ethical implications of emerging biotechnologies. Although the book is not about biotechnology or genomics per se, these domains are used as the substantive topics around which the different contributions explore various aspects of public engagement. The book is divided into six sections, each dealing with an aspect of the subject detailed below.

Part 1, "The Purpose and Function of Public Participation," gives an overview of democratic participation and its role in science and technology. The chapters by Gastil, Hennen, and Phillips provide a critical overview of participatory mechanisms from different perspectives. Gastil's focus is on the different models and approaches that grant lay citizens more direct means of influencing policy or legislation and that have been tried and have met with at least some success. Together with a historical overview of the jury as a citizen-deliberative body for judicial decision making, Gastil reviews these processes in the context of intercultural difference and then integrates them into an array of processes that could be used to work through cultural conflicts over emerging technology policy on issues such as biotechnology and genomics. In Chapter 2, Hennen draws on many years of experience in technology assessment (TA) in Europe to provide an overview of mechanisms of public participation in the regulation of emerging technologies used in that context. Using the history of the development of TA as a foundation, Hennen explores the tensions between experts and publics in developing regulatory frameworks. Finally, Phillips cautions us against assuming that democratic engagement is necessarily a solution to the problems that it espouses to solve. With a focus on the investments made by several OECD governments in engaging publics in various forms of consultation and collective decision making, Phillips critically examines the assertion that the move to engage the public is a strategic response to a perceived democratic deficit, compared with assertions that it is simply a tactical reaction to widespread controversy. Phillips argues that new norms of governance demand accountable, responsible, and transparent outcomes and that public engagement processes can enhance or undercut those goals, depending on how and when they are used.

Parts 2 and 3 deal with the issue of legitimacy in public participation. The chapters in these sections thus have to do with the implementation and positioning of deliberation with respect to external conditions (e.g., societal structures and values) and internal conditions (e.g., operational definitions for defining and achieving deliberation).

Following up on some of the challenges raised in the first section, in Part 2, "External Conditions for Legitimate Public Engagement: Ethics, Society, and Democracy," the focus is on the broader democratic frameworks in which participatory processes are embedded. Dodds, in particular, highlights the notion that engaging citizens in the governance of biotechnologies is not sufficient; effort needs to go into creating the conditions for democratic legitimacy that then provide a platform of trust on which meaningful public deliberation can rest. Dodds draws on work on deliberative democracy to unpack the links between pragmatist social epistemology and the political value and significance of participation in deliberation about emerging technologies and their governance. She argues that, without a significant level of well-founded public trust and substantive public accountability, public participation risks being misused as a diversion from substantive democratic justification. Gavaghan's chapter, accordingly, is dedicated to exploring an example of public participation in an area of biotechnology that he argues constitutes a misuse of public consultation. Focusing on the case of assisted reproductive technologies (ARTs) in the United Kingdom, New Zealand, and Canada, Gavaghan outlines the challenges posed to regulators by the ethical pluralism that surrounds many emerging biotechnologies. The chapter by Korthals concludes this section with an examination of the connection of particular deliberative exercises with their larger societal impacts. Drawing on Habermas and other deliberative theorists, Korthals considers such challenges as unequal participation in deliberations; cognitive and normative uncertainties in the life sciences; framing of issues; problems associated with multi-level governance; and the differential impacts that some scientific and technological projects seem to have compared with others.

Part 3, "Internal Conditions for Legitimate Public Engagement: Lessons for the Practitioner," focuses on the process of deliberation itself. The divide between theorists and practitioners of deliberation is an oft-recognized phenomenon. Cobb and O'Doherty attempt to bridge this gap by focusing on how particular public engagement projects need to be conceptualized, implemented, and analyzed such that they can be argued to represent a defensibly legitimate public voice on a given issue. Cobb's chapter engages with the question of how to define deliberation and how to measure the quality of its implementation. Pointing out that there is no commonly accepted definition of deliberation, Cobb begins by developing a working definition based on a review of the competing perspectives on deliberation. He then introduces the case of a novel study of citizen deliberation on human enhancement technologies and uses it to illustrate ways in which the outcomes of deliberation can be assessed, particularly in the context of typical criticisms of deliberation. The chapter by O'Doherty follows with an examination of how the conversations emerging from deliberation on the social and ethical implications of biotechnologies might be approached for the purpose of extracting legitimate conclusions. Arguing that this type of discourse is characterized by a distinct structure, O'Doherty develops the notion of *deliberative discourse.* If a given deliberative forum follows the normative guidelines for deliberation, then the resulting discourse should involve deliberants who are increasingly informed about the technical nature of the issues and should be increasingly inclusive of a broader range of perspectives and, ultimately, convergent toward a "group opinion." These observations have important consequences for the analysis of deliberations and the presentation of legitimate outcomes of deliberation to policy makers.

Part 4, "Institutional Contexts of Public Participation," builds on a theme touched on in other sections of the book, namely, that participation needs to be sensitive to socio-cultural and other contexts for successful implementation. In this section, closer attention is paid to the recognition that the particular institutional contexts into which specific technologies are launched in part define how publics need to be consulted. Appropriate public engagement needs to occur with a clear understanding of the relevant conditions and thus needs to be tailored rather than generic. In fact, certain institutional contexts can inhibit or even render public engagement altogether unsuitable for a given situation. The chapters by Castle, Weisbrot, and Migone and Howlett examine the emergence of different genomic technologies in particular settings. In each case, the authors pay close attention to the regulatory environment, the nature of the technology relative to the socio-cultural context, and the situated role of public participation. Castle does so in the context of nutrigenomics and the institutional context of a regulatory body, the Public Health Agency of Canada, and a professional body, Dieticians of Canada; Weisbrot focuses on the activities of the Australian Law Reform Commission and its public inquiries into the protection of human genetic information; and Migone and Howlett examine the role of public participation in the Canadian biotechnology regulatory regime. In all three cases, the authors highlight the critical approach necessary to ensure that public participation is not treated as an "off the shelf" solution but one that requires sensitivity and careful integration into existing frameworks.

Public participation in science and technology can be achieved through numerous avenues and mechanisms. While previous sections of the book focus on more traditional methods, such as citizen juries, deliberative forums, and parliamentary inquiries, Part 5, "Modes of and Experiments in Participation," considers different and more novel modalities that are available for engaging publics in emerging biotechnologies. In particular, Secko explores the use of journalism, Reid discusses drama-documentaries, and Danielson and Culver each focus on a different aspect of electronic and computer-mediated forms of engagement. Secko opens this section by exploring the role of journalism in public engagement in issues of science and technology, using the case of salmon genomics as a focus. He observes that, though journalism carries with it the intellectual seeds of a practical approach that can support democratic engagement initiatives, journalism as a theory and practice is as yet an underutilized tool in the realm of research-based citizen engagement. Secko's case study illustrates the potential of journalism to help generate the information required to deal with conceptually complex topics such as the ethical and social implications of biotechnology and genomics as well as to help reflect on and sustain such democratic debates. Reid's contribution is a case study of the BBC drama-documentary *If ... Cloning Could Cure Us* as a novel avenue for engaging publics in emerging biotechnologies. Reid focuses on criticisms of deficit models of public understanding of science, illustrating how a carefully crafted and interactive "dramadoc" goes beyond a deficit agenda in engaging viewers in the issue of therapeutic cloning.

In addition to the more traditional media of journalism and television, computer-mediated technologies have been hailed as having the potential to revolutionize possibilities for public engagement. The chapters by Danielson and Culver illustrate the complexity of considerations that need to be addressed if effective public engagement is to be achieved through these mechanisms. Danielson showcases a particular computer-mediated interface, the N-Reasons platform, and its use in engaging publics in emerging technologies. The platform is designed as an explicitly normative polling instrument to improve public participation in ethically significant social decisions. Danielson also outlines a normative theory that motivates claims that N-Reasons can improve *ethical* decision making and then shows empirically that decision making can be improved in this way. Culver casts a more critical glance at past failures of e-participation, focusing on his experience in the Canadian regulatory context and genetically modified organisms. Culver argues, however, that these failures are due less to an overrating of e-participation as a powerful mechanism for engaging publics than to the use of tools applied to the wrong kinds of problems. He concludes that the increasingly evident social demands of eco-innovation might be precisely the kind of domain in which e-participation will be of unique value.

A message that recurs throughout the book is that context matters. Both the design and implementation of public engagement need to be guided by consideration of the nature of the technology, relevant existing regulatory frameworks, and, perhaps most importantly, which stakeholders and publics are most affected. In Part 6, "Understanding Stakeholders and Publics," questions about the relationship between the particular technology, the relevant public(s), and the mechanism for engagement are examined in more detail. The chapters by Einsiedel, Plows, and Atkins each deal with these issues from a different perspective. Einsiedel's chapter is premised on the observation that much of the attention to public participation has been focused on participation by "mini-publics" engaged in structured deliberation that is typically organized by governments or academics. Einsiedel maintains that the dynamic processes of "contentious politics" occurring in public arenas and carried out by stakeholder groups are an important form of participation that provides another lens for understanding the social shaping of emerging technologies. Plows then takes on one of the themes highlighted by Einsiedel, namely, that public engagement can be motivated from without the policy sphere with very different implications for theory and practice. Focusing on a variety of publics engaging with human genetic technologies in the United Kingdom between 2003 and 2007, Plows explores public engagement as a form of social movement. She argues that, in addition to being a policy tool, public engagement can be understood as a network of relationships among individuals and groups, occurring in many different circumstances and taking many forms, but in which participants frame issues on their own terms. Finally, the chapter by Atkins provides a compelling personal account questioning whether the very liberal values that underlie so much of the inspiring reforms of the past 300 years still serve us in attempting to integrate emerging biotechnologies into our personal lives. In particular, Atkins questions whether liberal values facilitate moral decision making when they are applied to genomic and technological innovations. Paradoxically, she argues that liberalism actually encourages the pursuit of scientific and technocratic ends rather than the consideration of individual dignity. She concludes that liberal values, in this context, undermine moral considerations of respect for the individual rather than supporting them.

In summary, contemporary governance of emerging technologies reflects multi-faceted challenges that require complex accounts of how publics are engaged and participate. The public sphere for genomics and other emerging technologies offers arenas within which consultation, engagement, and participation have been deployed in various ways, and different publics have participated in diverse modes in the pursuit of varied social and policy goals. Although participation is not naively presumed to be a panacea, the purpose of this book is to bring together reflections on the complexities, lessons, and challenges of thinking through public participation as an important element in the governance of new and emerging technologies.

PART 1

THE PURPOSE AND FUNCTION OF PUBLIC PARTICIPATION

Giving Power to Public Voice
A Critical Review of Alternative Means of Infusing Citizen Deliberation with Legal Authority or Influence

JOHN GASTIL

On 26 June 2009, the governor of the State of Oregon signed a law that set up Citizens' Initiative Review (CIR) panels for the 2010 statewide general election. Twice in August 2010, a stratified random sample of twenty-four Oregon citizens convened to deliberate on a ballot initiative for five days; then they produced a written statement that went into the official voters' pamphlet sent to each registered voter by the secretary of state.[1] The CIR is unique in its use of officially sanctioned small-scale deliberation to improve the quality of large-scale electoral deliberation, and an initial assessment shows that the process had a clear impact on how voters understood and voted on the issues that the CIR studied (Gastil and Knobloch 2010). That assessment helped make the CIR a regular feature of Oregon elections, since the legislature and governor made the process permanent in 2011.

Oregon has showcased just the latest in a series of recent innovations in citizen deliberation. The past two decades have seen the development and advocacy of specific deliberative reforms (Fishkin 2009; Gastil and Levine 2005; Goodin 2008; Nabatchi et al. 2010) that have built on projects that began in the 1970s and 1980s (Becker and Slaton 2000; Crosby 1995; Hendriks 2005). Although the bulk of deliberative theory originally developed in the United States and Europe (Chambers 2003), some of the most important deliberative processes have sprung up in Canada (Warren and Pearse 2008), Brazil (Coelho et al. 2005; Wampler 2007), and India (Fischer 2006).

This chapter's primary purpose is to review some of these processes and suggest ways that they can be adapted and improved to create an array of deliberative designs appropriate for setting policy on emerging technologies, such as biotechnology and genomics. The chapter has a second purpose. The discussion of deliberation proceeds against the backdrop of a particular conception of intercultural difference. I intend to suggest precisely how different models of deliberation might tackle the problem of cultural differences, which can pose a considerable challenge to the idea of deliberating to reach a collective judgment, let alone a public consensus. It is with this question of intercultural communication that I begin.

The Intercultural Challenge

There are innumerable ways of conceptualizing culture, and here I use one: the two-dimensional cultural typology employed by the Cultural Cognition Project (see www.culturalcognition.net), an effort to systematize this approach in regard to how culture shapes attitudes toward risk and public policy as well as deliberation on those subjects. The basic idea is that each person has a particular cultural orientation (Kahan et al. 2007; Wildavsky 1987). Within any given society, individuals vary in the degree to which they believe society should be organized along individualist versus collectivist principles and between egalitarian and more hierarchical/traditional principles.

One's culture is influenced but not determined by the character of one's family, community, or geographic region of origin, and one's biological sex, ethnicity, and socio-economic status can combine to make a given cultural mode more or less desirable. In any case, the emphasis here is not the *origins* of one's cultural orientation so much as its *consequences* – particularly when thinking about designing a deliberative process that will include people who hail from different orientations.

Cultural orientations provide the engine that drives selective attention to information (Mutz and Martin 2002), message framing (Scheufele 1999), filtering (Zaller 1992), and heuristic processing of information (Lupia and McCubbins 1998). Thus, culture plays a pivotal role in shaping attitudes, and, if one hopes to create a deliberative process for examining issues such as emerging technologies, these different cultural orientations pose two related challenges.

The first challenge is that citizens approach even new technologies with a pre-existing set of cultural biases – often in conflict with one another. Each cultural orientation has a different starting point regarding science and innovation (Gastil et al. 2008). Hierarchs begin with skepticism about any enterprise that grows out of liberal university systems, and they couple this doubt with general trepidation about deviations from the status quo, including technological change. Egalitarians start with a different skepticism – this time of corporate-sponsored research and development and the techo-dystopia that it might induce; however, egalitarians also view innovation as – at least potentially – liberating, as they alleviate the drudgery and suffering of society's least fortunate members. Along the other cultural dimension, individualists celebrate scientific discovery and technological entrepreneurialism, with reservations only about its use for (or association with) surveillance and social control. Their counterparts, the collectivists, are leery of science and technology when it breaks up a social consensus and weakens community ties.

The second and greater challenge is in getting people from different orientations to take seriously the varied points of view that they hear *during* the deliberation. In effect, any short-term deliberative process is up against a life-long backdrop of cultural cueing and attitude formation. Even with new technologies, people already have general biases that conflict across cultural lines. Research has shown that more pronounced cultural divergence occurs quickly even when people are given neutral information about new phenomena, such as nanotechnology (Kahan et al. 2008) or the human papillomavirus (HPV) vaccine (Kahan et al. 2010).

At this stage in the Cultural Cognition Project's development, the ideas for reaching across such cultural divides remain theoretical (Gastil et al. 2008; Kahan et al. 2006), but the general approaches are as follows.

- Encourage the recognition of different cultural orientations and affirm the distinct value commitments underlying them.
- Introduce basic factual information but recognize the difference between raw empirical evidence and the *cultural interpretation* of that evidence (e.g., its weight, implications).
- Point out the variation in policy positions *within* the different orientations; whenever possible, find advocates from opposite orientations who argue for the same policy choice.

If each of these succeeds, then one will have created an environment in which people from different orientations feel respected, can express their views freely, and can weigh information and alternative solutions together fairly and thoughtfully. Such a process is often called "deliberative" (Burkhalter, Gastil, and Kelshaw 2002; Gastil 2008), and it is to the design of such processes that I now turn.

Authoritative/Influential Deliberative Methods

In reviewing the variety of available deliberative methods, I focus on the smaller subset of processes that either has a degree of legal authority or has a reasonable expectation of direct influence on policy.[2] Table 1.1 presents a set of eight models of deliberation, each of which has either a legal authority or a clear path of influence.

Overview of the Eight Processes

First on the list is the jury system – the most venerable existing deliberative institution. Although overlooked in most discussions of public deliberation (Gastil 2008), the voluminous research on this method of public involvement reminds us of the ability of jurors to reach sound verdicts, even in complex civil cases (Vidmar and Hans 2007). Research also shows the civic impact that jury service has on the everyday citizens who take seriously their charge to resolve legal questions on behalf of the state (Gastil et al. 2010). The power of the jury is tremendous, even though judges can narrowly define its duties, constrain its knowledge, and consider appeals of its verdicts. Although much maligned in popular media, in those countries where it is common practice, the jury has essentially *defined* the term "deliberation" for the general public.

A far more recent and academically celebrated process is participatory budgeting, first developed in Brazil, along with a raft of other participatory processes (Coelho et al. 2005; Wampler 2007). In a nutshell, this process gives average citizens (and/or their surrogates embodied in non-governmental organizations [NGOs]) a stronger voice in local decision making. In Brazil, this process has created a more vibrant local public sphere in which everyday citizens can become politically sophisticated and influential to a degree that was previously implausible (Baiocchi 2003). Outside Brazil, this process has sometimes been influential without having legal authority (Cabannes 2004), but it has certainly added to the popular imagination the idea that everyday citizens could exercise substantial budgetary authority responsibly and thoughtfully.

A related process that has received less notice is the people's campaign undertaken in the distinctive southwestern Indian state of Kerala – the site

TABLE 1.1

Comparison of eight methods of authoritative/influential public deliberation

Deliberative process	Initial location	Inception	Participation	Authority/influence	Legislative/executive/judicial check
Civil/criminal jury	London, England	1670*	6-12 via random selection (and usually *voir dire*)	Resolve matters of legal dispute	On specific points of law, judgments can be appealed to higher courts
Participatory budgeting	Porto Alegre, Brazil	1989	100s/1,000s open, then elect delegates	Prioritize and allocate portion of city budget	Mayor can veto
People's campaign	Kerala, India	1996	100s/1,000s open, then elect delegates	Decide how to spend a third of state planning budget	Legislative vote required to authorize spending
Deliberative poll	Wenling City, China	2005	235 via random selection	Set priorities for economic development	Government action needed to implement priorities
Biobank deliberation	British Columbia, Canada	2007	21 via stratified random selection	Consensus recommendations and points of disagreement	Government action needed to act on recommendations and resolve disagreements
Citizens' parliament	Canberra, Australia	2009	150 via stratified random selection	Draft and prioritize political reforms given directly to federal government	Government action needed to implement priorities
Citizens' assembly	British Columbia, Canada	2003	160 via random selection (+ 2 Aboriginal)	Send electoral reform referendum to electorate	Legislative vote required to trigger referendum
Citizens' initiative review	Oregon, USA	2010	24 via stratified random selection	Present analyses and recommendations through the voters' guide	Secretary of state oversight (could be replaced by a board consisting of former panelists and state appointees)

* There is no definitive date at which the "modern civil jury" became established. This particular date is the year of "Bushel's Case," which established that juries could not be punished by the judge for rendering a "not guilty" verdict (Vidmar and Hans 2007, 27).

of many democratic innovations (Fischer 2006). Like participatory budgeting, the people's campaign draws in the general public through local NGOs and other social networks, and these large assemblies discuss local issues and send representative delegates to work out details. The people's campaign goes beyond budgeting per se to consider wider questions of economic development, and it helps to model ways of drawing large populations into a deliberative process that can still be channelled down to smaller bodies of discussants capable of reaching agreements that carry considerable popular legitimacy.

Unlike the preceding processes, the deliberative poll has usually been strictly advisory, but one important exception was its use in China in 2005, where the poll results used to set local priorities for a public works project (Fishkin 2009; Leib and He 2006). The key difference between this and the Brazilian and Indian examples was that the participants were a large, randomly selected body that deliberated face to face and then recorded its preferences privately, as individuals, through a survey. This method likely reduces – for better or worse – the potential for broader social influence on final choices, so it models another important alternative for recording and aggregating citizens' informed judgments.

The biobank deliberation stands apart from the others described thus far in that it was a process aimed at finding not only common ground but also key points of disagreement (O'Doherty and Burgess 2008). This project divided a relatively small group of twenty-one participants into separate subgroups to consider ethical issues concerning the development of biobanks for research. This method – analogous to a common group procedure called dialectical inquiry (Schwenk 1989) – yielded stronger small-group agreement than large-group consensus. For the purpose of this chapter, however, that dissensus was the most interesting feature, and the biobank example demonstrates the potential for deliberation to clarify points of disagreement that can be fruitful to explore in later deliberative stages.

Like the biobank deliberation, the Citizens' Parliament held in Australia in 2009 (Dryzek 2009) lacked formal authority and hoped only to influence government policy making. One of the distinctive features of its design was its emphasis on prioritization rather than exclusively on decision making (on similar processes, see Carson and Hartz-Karp 2005; Lukensmeyer, Goldman, and Brigham 2005). Working with a large body of people over the course of four days, the parliament aspired to work through concrete proposals for political reform – but only to the point that the participants could evaluate and rank each in terms of different criteria. Again, such an event seems to have a useful role as part of a larger deliberative process.

The British Columbia Citizens' Assembly (Warren and Pearse 2008) had a more demanding task: it was designed to advance a concrete proposal for revising the voting system in British Columbia. It did weigh alternatives, but ultimately, it had to make a clear choice – a recommendation spelled out in sufficient detail that it could be put to a vote of the full provincial electorate. Although the assembly's proposal ultimately won support from a majority of voters, it failed to reach the 60 percent threshold required for passage. A revote held a few years later failed to even win a majority.[3] The precedent was established, though, that a body of deliberative citizens could create credible legislation through a focused meeting process lasting months.

The final case considered here is the CIR process noted at the outset of this chapter.[4] Like the citizens' assembly, this process is interfaced with a larger voting public, but rather than drafting a law and forcefully recommending it, the CIR evaluates laws proposed *by others* through the referendum/initiative process, and it produces a statement that provides neutral analysis as well as key pro and con arguments. Moreover, whereas the citizens' assembly had an ambiguous, underfunded, public outreach component, the CIR has a prominent and powerful place in the election – a full page in the official Oregon voters' pamphlet distributed to every voting household. Research on the 2010 CIR showed that, as intended, it had a significant impact on the voting choices made by the wider electorate (Gastill and Knobloch 2010).

Agenda Setting, Decision Making, and Cultural Difference

When comparing the authority and influence of these processes, it is important to look at a few key features, which I have juxtaposed in Table 1.2. Beginning with the left-hand column, it is noteworthy that *none* of these processes empowers citizens to set the agenda – that is, to determine the scope and policy focus of their deliberations. Next, though I refer to each as being deliberative, the analytic purposes of these bodies vary tremendously, with the most common role being prioritizing different ideas and interests. Only the jury stresses the importance of a final judgment, and the biobank and CIR stand out as the processes oriented toward generating (clarifying, really) debate. To the extent that there is a decision rule explicitly established, it is majoritarian – with only the jury formally requiring a supermajority. Finally, there is no direct policy-voting role for the full public beyond those

TABLE 1.2

Agenda setting, analysis, and role for popular voting in eight methods of public deliberation

Deliberative process	Agenda setter	Analytic function	Decision rule	Role for a popular vote*
Civil/ criminal jury	Litigants/state	Judgment	Supermajority or unanimity	None
Participatory budgeting	Government/ NGO	Prioritization	Majority	Elect delegates
People's campaign	Government/ NGO	Prioritization	Majority	Elect delegates
Deliberative poll	Government/ convenor	Prioritization	Implicit plurality	None
Biobank deliberation	Convenor	Recommendations and debate	Implicit plurality	None
Citizens' parliament	Government	Prioritization	Implicit plurality	None
Citizens' assembly	Convenor/ participants	Policy analysis and policy design	Majority	Directive to public referendum
Citizens' initiative review	Initiative proponents	Policy analysis	Majority	Advisory to public referendum

* In all of the countries where these systems are employed (except China), the election and re-election of public officials who oversee, convene, and/or implement policies developed by these deliberative bodies are subject to a popular vote. That is a non-trivial general public role, but it is not directly related to the deliberative bodies per se.

who are selected at random, with the exceptions of the citizens' assembly and the CIR, which link to – but do not directly involve – the vote of a wider public. Thus, none of these deliberative processes aspires to full public judgment; rather, they simply present to the wider public (or rely entirely on) the judgment of a representative or mobilized subset of the public.

Relating these features back to the aforementioned cultural theory, two things become apparent. First, there is a clear optimism underlying each process that deliberation can discover a majority (or plurality) by bringing together culturally diverse publics and giving them a specific agenda and analytic purpose. If we look more closely at the procedural directives in each process, we will find wide variation from the limited deliberation instructions given to juries to the question-and-answer orientation of deliberative polling to the more intensive discussion methods of the biobank and CIR panels (see Crosby 1995 for the citizen jury model that influenced both). One can take the view that any of those will work, owing to the natural power of even unstructured deliberation, or (more plausibly) one can argue that effective cross-cultural deliberation will occur as long as the process has a degree of gravity, plus an emphasis on honest and respectful discussion. Either way, the first assumption remains that deliberation *can* generate plurality – or even supermajority – positions across lines of cultural difference.

Second, there is an implicit assumption that the larger public will *accept* the judgment of its deliberative peers, even without having participated directly in the process. This optimism presumes either that the deliberative body – or its representatives – can convey the ideas or symbolic content that permits a cross-cultural agreement or that the public will simply accept the consensus claims made by the body or representatives, owing to the credibility they had pre-established or won through their intensive deliberation.

There exists no sustained research to discern whether either of those assumptions is, in fact, warranted, and the cultural cognition research referenced earlier should give one pause. As I move into the next section, where I suggest an integrated set of deliberative procedures, I attempt to highlight which limited aims each process can more reasonably be expected to produce by way of intercultural understanding or agreement.

Deliberative Design Modifications/Integrations for Technology Policy
Although the preceding set of eight deliberative designs is by no means exhaustive, it does provide plenty of raw material from which one can craft modified or integrated designs. Here I do precisely that, with an eye toward how each new method can complement the others and address the cross-cultural problem from a different angle. By way of tabular summary – clearly the presentational partner to which this chapter has wedded itself – Table 1.3 shows the five deliberative methods that I wish to propose.

Policy Jury
The policy jury would draw on the civil/criminal jury model to create a body capable of handling case-specific disputes. It might seem ironic to use juries as a kind of "alternative dispute resolution," since one often thinks of juries (at least in the American context) as the very process to which mediation or

TABLE 1.3

Distinct roles for five different methods of deliberative public decision making on emerging technologies

Deliberative process	Design based on ...	Function/authority	Special features	Cross-cultural aspiration
(1) Policy jury	Civil/criminal jury	Resolves individual extralegal disputes and problematic cases	"Jury of peers" can represent special populations (e.g., patients); uses a five-sixths majority rule with twelve jurors	Ask jurors to reach a culturally neutral consensus based on limited evidence and legal issues
(2) Participatory technology campaign	Participatory budgeting and people's campaign	Agenda setting for future deliberation: which questions should we put before the public?	Gives relevant stakeholder NGOs and general public a clear role in setting up deliberation; majority rule	Encourage a superordinate identity as community member to transcend cultural differences
(3) Question framing panel	Biobank deliberation	Frames the question for deliberation, highlighting key areas of disagreement	Intensive face-to-face deliberation in a small group with no consensus pressure	Recognize and draw out cultural differences in finding points of disagreement
(4) Special advisory parliament	Deliberative poll and citizens' parliament	Gives a clear indication of the balance of deliberative judgment on a policy question	Deliberation focuses on a clear question and addresses likely points of disagreement	Address cultural conflicts head on to discern which might give way and which cannot
(5) Referendum assembly	Citizens' assembly and citizens' initiative review	Drafts specific policy question to be put to a vote of a large electorate (e.g., state/provincial/federal)	Also prepares a voters' guide (sent to every household), which includes a consensus statement and the key pro and con arguments raised	Identify points of transcultural consensus and send out clear cultural cues to voters (via majority and minority statements)

arbitration is the alternative. Setting up juries with the authority to resolve cases, however, might be appropriate for cases that *would not* directly end up in the courts. For instance, if a new genomics research campus were all but complete, save for a final ethical dispute among its organizers and critics, it might be appropriate to toss that final question to a well-structured policy jury that would have the power to settle the matter.

In terms of cultural conflict, the jury design is meant to narrow the focus and relevant considerations to the point that jurors might be able to play the role of impartial judges. The standard jury is conceived of as a body that is given a factual question in a specific legal context and that aspires to a neutral judgment, taking into account the broadest influences of one's life experience – hence the desire for a jury of "peers." Numerous mock jury studies have found, nonetheless, evidence of how verdicts can be influenced by jurors' backgrounds – including their cultural orientations (Kahan 2010; Kahan, Hoffman, and Braman 2009). Yet, in actual practice, demographic and attitudinal effects on jury verdicts are very small (Lieberman and Sales 2006), so, for more narrowly defined deliberative challenges, this might be an appropriate model for reaching a cross-cultural judgment.

Participatory Technology Campaign

The remaining models that I advance all presume that a policy jury would not be sufficient for the policy challenge at hand. The first of these, the participatory technology campaign, tries to harness the energy and interests of a wider civil society to set the agenda for future deliberation. As noted earlier, none of the cases reviewed here gave the general public a definitive role in agenda setting, even within a particular policy domain, such as biotechnology. Typically, government, interest groups, academics, or civic associations select what goes into the deliberative agenda, and this process aims to present an alternative. If there is a weakness in the participatory budgeting and people's campaign methods, it might be the prominent role given to organizations – rather than individual citizens – but when repurposed in this way, I think that it gives those entities an appropriate role. Civic and interest-based organizations form to help articulate and advocate public concerns, and this forum would give them – and their citizen members – the chance to prioritize among the various issues that they have identified.

If the groups and individuals who join the participatory technology campaign represent a wide-enough cross-section of the relevant political unit (city, state, nation), then the campaign can promote a "superordinate identity" to transcend cultural differences (Gaertner et al. 1999; Lee 2005).

Successfully creating and maintaining such an identity might be tenuous in a large-scale deliberation that has to reach a definitive policy conclusion, but the stakes are slightly lower for the campaign: its job is simply to *prioritize* issues for other deliberative processes to resolve.

Question Framing Panel

With an issue in hand, one might want to deploy a question framing panel, which I conceive of as following the model of the biobank deliberation. This requires only a small group of randomly chosen citizens who are actively encouraged – whether in smaller groupings or not – to both find common ground and, more importantly, discover their key areas of disagreement. Because this group would have no pressure to arrive at a consensus, it would easily find and clarify substantive and symbolic divisions, be they cultural or otherwise. This panel would face the least daunting cultural task since it would ask for cross-cultural consensus only on those points on which it comes easily.

Special Advisory Parliament

If an issue has been selected by a participatory technology campaign and framed by a question framing panel, then it is ripe for the next process – a special advisory parliament. This body would bring together a hundred or more randomly selected citizens to work through the issues and controversies already discovered in the previous stages. In the end, this process would offer a clear indicator of the balance of deliberative judgment on a policy question. By recognizing points of disagreement at the outset, it would be understood that such a body might not reach consensus; at the same time, there would be a measure of social pressure to work through those clearly defined conflicts, and some might fall away in the course of deliberation.

Of all the cross-cultural aspirations, this is the greatest. Success would depend on effectively deploying each of the three strategies identified earlier. First, framing the issue would need to openly recognize the divergent perspectives of different cultural orientations on the issue, highlighting not just substantive but also value conflicts therein. Second, the factual information brought forward would also likely acknowledge different cultural interpretations thereof, not to promote an empirical relativism but to encourage understanding and working through those interpretive conflicts. Third, expert panels used in this process should put forward advocates from opposite cultural orientations who argue for the same policy choice, such that there is a pluralism of perspectives both within and between the different cultural orientations. This will give the citizen participants the freedom to explore their own views without feeling disloyal to their own particular cultural worldviews.

Referendum Assembly

If one wanted more than an advisory parliament, one could empower a comparable body to meet over a longer period of time and develop a recommendation to put to a public vote. Most closely modelled on the citizens' assembly, this body would have one additional power: it would draft a formal recommendation that would be mailed to every registered voter, including key findings and pro and con statements, as done in the CIR. This process aspires to reach a cross-cultural agreement, but it allows a voice for any viewpoint that ends up as a minority perspective. That feature might weaken the force of this body's recommendation, but it will also reduce the chance of its majority view's silencing dissent.

Conclusion

To be clear, the preceding deliberative designs do not pretend to brush away cultural conflicts, nor do they constitute revolutionary advances in the way that we theorize and practice deliberation. Rather, they are meant simply to advance how we think about cultural difference, how we translate those differences into deliberative challenges, and which designs we might deploy to mitigate or manage those problems.

In a given policy context, it is likely that one would draw on only one or two of these deliberative processes, but that is the point – to think about one's particular policy objectives, the nature of the cultural conflict in play, and the appropriate role for public deliberation, be it to identify, frame, analyze, or resolve a debate. The choice that one makes might also reflect the role that the wider public wants to play – from putting an issue onto the deliberative agenda to voting on a deliberative recommendation. In my view, a mature public participation agenda would turn to different processes at different times, always with an eye toward what is the best use of the public's limited time and capacity for cross-cultural consensus.

NOTES

1 For the complete text of this bill, see http://gov.oregonlive.com/. Descriptions of the CIR process and scheduled dates come from the organization Healthy Democracy

Oregon, which championed the CIR through the Oregon legislature and served as the organizer of the panels. For more information on the CIR, see Gastil and Knobloch (2010).

2 There exists a vast array of other, more strictly advisory, methods of public engagement, consultation, and deliberation (Bingham, Nabatchi, and O'Leary 2005; Gastil and Levine 2005), and there are important approaches that emphasize collective action at least as much as collective judgment (Leighninger 2006).

3 For a description and detailed results, see http://www.elections.bc.ca/.

4 For earlier writings on this idea, see Crosby (2003) and Gastil (2000).

REFERENCES

Baiocchi, G. 2003. "Emergent Public Spheres: Talking Politics in Participatory Governance." *American Sociological Review* 68: 52-74.

Becker, T., and C.D. Slaton. 2000. *The Future of Teledemocracy.* New York: Praeger.

Bingham, L.B., T. Nabatchi, and R. O'Leary. 2005. "The New Governance: Practices and Processes for Stakeholder and Citizen Participation in the Work of Government." *Public Administration Review* 65: 547-58.

Burkhalter, S., J. Gastil, and T. Kelshaw. 2002. "A Conceptual Definition and Theoretical Model of Public Deliberation in Small Face-to-Face Groups." *Communication Theory* 12: 398-422.

Cabannes, Y. 2004. "Participatory Budgeting: A Significant Contribution to Participatory Democracy." *Environment and Urbanization* 16: 27-46.

Carson, L., and J. Hartz-Karp. 2005. "Adapting and Combining Deliberative Designs: Juries, Polls, and Forums." In *The Deliberative Democracy Handbook: Strategies for Effective Civic Engagement in the Twenty-First Century,* edited by J. Gastil and P. Levine, 120-38. San Francisco: Jossey-Bass.

Chambers, S. 2003. "Deliberative Democratic Theory." *Annual Review of Political Science* 6: 307-26.

Coelho, V., P. Schattan, B. Pozzoni, and M. Cifuentes Montoya. 2005. "Participation and Public Policies in Brazil." In *The Deliberative Democracy Handbook: Strategies for Effective Civic Engagement in the Twenty-First Century,* edited by J. Gastil and P. Levine, 174-84. San Francisco: Jossey-Bass.

Crosby, N. 1995. "Citizen Juries: One Solution for Difficult Environmental Questions." In *Fairness and Competence in Citizen Participation: Evaluating Models for Environmental Discourse,* edited by O. Renn, T. Webler, and P. Wiedemann, 157-74. Boston: Kluwer Academic Publishers.

–. 2003. *Healthy Democracy: Bringing Trustworthy Information to the Voters of America.* Minneapolis: Beaver's Pond.

Dryzek, J. 2009. "The Australian Citizens' Parliament: A World First." *Journal of Public Deliberation* 5. http://services.bepress.com/.

Fischer, F. 2006. "Participatory Governance as Deliberative Empowerment: The Cultural Politics of Discursive Space." *American Review of Public Administration* 36: 19-40.

Fishkin, J.S. 2009. *When the People Speak: Deliberative Democracy and Public Consultation.* Oxford: Oxford University Press.

Gaertner, S.L., J.F. Dovidio, J.A. Nier, C.M. Ward, and B.S. Banker. 1999. "Across Cultural Divides: The Value of a Superordinate Identity." In *Cultural Divides: Understanding and Overcoming Group Conflict,* edited by D.A. Prentice and D.T. Miller, 173-212. New York: Russell Sage.

Gastil, J. 2000. *By Popular Demand: Revitalizing Representative Democracy through Deliberative Elections.* Berkeley: University of California Press.

—. 2008. *Political Communication and Deliberation.* Thousand Oaks, CA: Sage.

Gastil, J.E., P. Deess, P.J. Weiser, and C. Simmons. 2010. *The Jury and Democracy: How Jury Deliberation Promotes Civic Engagement and Political Participation.* New York: Oxford University Press.

Gastil, J., and K. Knobloch. 2010. *Evaluation Report to the Oregon State Legislature on the 2010 Oregon Citizens' Initiative Review.* http://faculty.washington.edu/.

Gastil, J., and P. Levine, eds. 2005. *The Deliberative Democracy Handbook: Strategies for Effective Civic Engagement in the Twenty-First Century.* San Francisco: Jossey Bass.

Gastil, J., J. Reedy, D. Braman, and D.M. Kahan. 2008. "Deliberation across the Cultural Divide: Assessing the Potential for Reconciling Conflicting Cultural Orientations to Reproductive Technology." *George Washington Law Review* 76: 1772-97.

Goodin, R.E. 2008. *Innovating Democracy: Democratic Theory and Practice after the Deliberative Turn.* New York: Oxford University Press.

Hendriks, C.M. 2005. "Consensus Conferences and Planning Cells: Lay Citizen Deliberations." In *The Deliberative Democracy Handbook: Strategies for Effective Civic Engagement in the Twenty-First Century,* edited by J. Gastil and P. Levine, 80-110. San Francisco: Jossey-Bass.

Kahan, D.M. 2010. "Culture, Cognition, and Consent: Who Perceives What, and Why, in 'Acquaintance Rape' Cases." *University of Pennsylvania Law Review* 158: 729-812.

Kahan, D.M., D. Braman, G.L. Cohen, J. Gastil, and P. Slovic. 2010. "Who Fears the HPV Vaccine, Who Doesn't, and Why? An Experimental Study of the Mechanisms of Cultural Cognition." *Law and Human Behavior* 34: 501-16.

Kahan, D.M., D. Braman, J. Gastil, and P. Slovic. 2007. "Culture and Identity-Protective Cognition: Explaining the White-Male Effect in Risk Perception." *Journal of Empirical Legal Studies* 4: 465-505.

Kahan, D.M., D. Braman, P. Slovic, J. Gastil, and G. Cohen. 2008. "Cultural Cognition of Nanotechnology Risk-Benefit Perceptions." *Nature Nanotechnology* 4: 87-90. http://www.nature.com/.

Kahan, D.M., D.A. Hoffman, and D. Braman. 2009. "Whose Eyes Are You Going to Believe? *Scott v. Harris* and the Perils of Cognitive Illiberalism." *Harvard Law Review* 122: 837-906.

Kahan, D.M., P. Slovic, D. Braman, and J. Gastil. 2006. "Fear and Democracy: A Cultural Evaluation of Sunstein on Risk." *Harvard Law Review* 119: 1071-1109.

Lee, S. 2005. "Judgment of Ingroups and Outgroups in Intra- and Intercultural Negotiation: The Role of Interdependent Self-Construal in Judgment Timing." *Group Decision and Negotiation* 14: 43-62.

Leib, E.J., and B. He, eds. 2006. *The Search for Deliberative Democracy in China.* New York: Palgrave Macmillan.

Leighninger, M. 2006. *The Next Form of Democracy: How Expert Rule Is Giving Way to Shared Governance.* Nashville: Vanderbilt University Press.

Lieberman, J.D., and B.D. Sales. 2006. *Scientific Jury Selection.* Washington, DC: American Psychological Association.

Lukensmeyer, C.J., J. Goldman, and S. Brigham. 2005. "A Town Meeting for the Twenty-First Century." In *The Deliberative Democracy Handbook: Strategies for Effective Civic Engagement in the Twenty-First Century,* edited by J. Gastil and P. Levine, 154-63. San Francisco: Jossey-Bass.

Lupia, A., and M.D. McCubbins. 1998. *The Democratic Dilemma: Can Citizens Learn What They Need to Know?* New York: Cambridge University Press.

Mutz, D.C., and P.S. Martin. 2002. "Facilitating Communication across Lines of Political Difference: The Role of Mass Media." *American Political Science Review* 95: 97-114.

Nabatchi, T., J. Gastil, M. Weiksner, and M. Leighninger, eds. 2012. *Democracy in Motion: Evaluating the Practice and Impact of Deliberative Civic Engagement.* New York: Oxford University Press.

O'Doherty, K.C., and M.M. Burgess. 2008. "Engaging the Public on Biobanks: Outcomes of the BC Biobank Deliberation." *Public Health Genomics* 12: 203-15.

Scheufele, D.A. 1999. "Framing as a Theory of Media Effects." *Journal of Communication* 49: 103-22.

Schwenk, C. 1989. "A Meta-Analysis on the Comparative Effectiveness of Devil's Advocacy and Dialectical Inquiry." *Strategic Management Journal* 10: 303-6.

Vidmar, N., and V.P. Hans. 2007. *American Juries: The Verdict.* Amherst, NY: Prometheus.

Wampler, B. 2007. *Participatory Budgeting in Brazil: Contestation, Cooperation, and Accountability.* University Park: Pennsylvania State University Press.

Warren, M., and H. Pearse. 2008. *Designing Deliberative Democracy: The British Columbia Citizens' Assembly.* Cambridge, UK: Cambridge University Press.

Wildavsky, A. 1987. "Choosing Preferences by Constructing Institutions: A Cultural Theory of Preference Formation." *American Political Science Review* 81: 3-21.

Zaller, J.R. 1992. *The Nature and Origins of Mass Opinion.* Cambridge, UK: Cambridge University Press.

Parliamentary Technology Assessment in Europe and the Role of Public Participation

LEONHARD HENNEN

27-44

The need for more scientific expertise in decision making and the fact that science and technology (S&T) are no longer uncontested in society brought the idea of technology assessment (TA) into being. Consequently, it is not surprising that the discourse on the concepts and methods of TA has always circled around the questions of how and which expertise has to be employed in TA and how (and to what extent) social interests or the view of the public must be included in the process. These questions are closely tied to the two impulses that have always driven TA (Guston and Bimber 2000): one drives toward expert analysis, while the other drives toward public deliberation. Accordingly, two models of TA have been pursued throughout its history: a policy analysis model and a public deliberation model.

When the Office of Technology Assessment (OTA) at the US Congress was established in the early 1970s, the policy analysis model was predominant. Stakeholders and public interest groups have, nonetheless, always played an important role in OTA studies. The deliberation model gained importance in Europe during the 1980s and 1990s and nowadays can be regarded as dominant in many European countries. This fact has to be seen in the context of major shifts in the relationship between science and society, shifts that are reflected in the social sciences in terms such as "reflexive modernization" (Beck, Giddens, and Lash 1994) and "mode 2 knowledge production" (Nowotny, Scott, and Gibbons 2001). The participatory turn can thus be understood as a reaction to specific difficulties in dealing with uncertainties or ambiguities in practical, normative, and ethical problems that inevitably arise from technical modernization and raise issues regarding the legitimacy of policy making (Hennen 1999).

After two decades of experience in participatory TA (hence PTA) in Europe, we are currently facing skeptical reflections on the epistemological and democratic roles and functions of participation in S&T policy making. More research is needed to clarify the civic epistemologies of participatory procedures as well as their relations to established procedures of representative democracy.

Parliamentary TA and Participation in Europe

When the OTA was closed, TA – as an import from the United States – had already become a major success in Europe. Today, the European Parliamentary Technology Assessment Network (EPTA, www.eptanetwork.org) comprises fourteen national and regional parliamentary TA institutions, the TA body of the European Parliament, and another four associate members with working relationships with their national parliaments. Parliamentary TA in Europe took up the heritage of the OTA but differs from it in terms of organization, methodology, and mission (Vig and Paschen 2000). Different institutional models are being followed in different countries, depending on their political or parliamentary traditions and cultures.

In some countries (e.g., Italy, Finland, and Greece), parliamentary TA committees have been established and (according to their agendas) invite experts to meetings or organize workshops and conferences to obtain scientific input to their debates. In the case of France, the individual members of the committee carry out TA studies on their own and deliver the results in the form of reports to their parliament. In other countries, parliaments have chosen a model of institutionalization that is closer to the OTA type. The parliament then runs a scientific office on a contract basis with a scientific institute (e.g., in Germany and at the European Parliament) or as part of the parliamentary administration (e.g., in the United Kingdom) to which TA studies are commissioned according to the information needs of the parliament. These studies can result in short parliamentary briefing notes or in full-fledged TA reports that draw both on in-house research and on input from a number of external scientific experts and stakeholders. A third type of parliamentary TA body is characterized by close cooperation between the parliament and external independent institutes (in some cases related to the national academies of sciences), which support parliamentary deliberations

FIGURE 2.1

The intermediate role of parliamentary TA in Europe

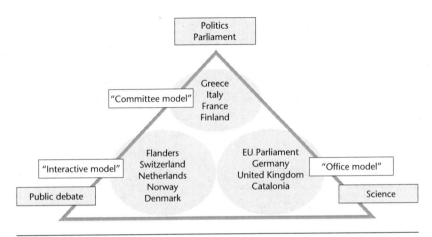

by providing policy reports or organizing workshops or hearings. Often this kind of arrangement involves an additional – interactive – mission of the institute, which opens the classical (OTA-like) TA setting of experts and policy makers to an additional third party, the general public. The mission of TA is then not only to support politics by providing in-depth and unbiased analysis of the possible effects of science and technology on society but also to inform and intervene in public debates (e.g., in Denmark, the Netherlands, Switzerland, Flanders, and Norway). This kind of orientation of the consulting process toward the public, stakeholders, societal groups, and citizens can be regarded as the European "improvement" on the classical TA model. The public is involved not only as an object of research but also as an actor in the TA process in its own right.

Against the background of these three models of institutionalization – the "committee," the "office," and the "interactive" models – it has to be stressed that, independent of the model applied, TA always plays an intermediate role with regard to three societal arenas: science, politics, and the public sphere (see Figure 2.1). Any TA institution has to position itself in this triangle. And even the more "classical" model of scientific policy consulting does not imply a closed circle-type of policy advice in which experts and policy makers negotiate behind closed doors. The TA process must always be transparent to the general public – in particular for social groups that have stakes in the issues to be addressed. On the other hand, TA that has a focus on intervention in public debates – for example by organizing citizen conferences, setting up lay panels, or holding public stakeholder meetings – cannot function without the backing of independent scientific expertise and will be politically meaningless without the involvement of the relevant policy-making bodies.

Recognition that the comprehensive assessment of new technologies is dependent on the values and interests of social groups has led TA to try out a large number of participatory formats for evaluating technology in which experts, laypersons, and political decision makers cooperate in various ways. Without going into detail about the various participatory methods, one can show that the three models of institutionalization of TA have a bearing on the application and status of the methods employed to involve stakeholders and the general public.

The involvement of stakeholders and the public might take the form of social research. The task of "social mapping" – that is, describing the societal interests and values involved in the issue at stake as well as societal debate and conflict – is an indispensable aspect of every TA project. In a setting dominated by the office model (as in the OTA and nowadays some TA institutions in Europe), the methodologies for involving the public range from the classical methodologies of social research (e.g., opinion polls or focus groups) to workshops with representatives of stakeholder groups. In any case, the methods applied do not involve any attempt to have a direct effect on the public or policy making; rather, they survey interests and values that might be relevant for policy making and thus have to be taken into account in policy analysis and recommendations to policy makers. The role of the public can vary from sheer objects of research, as in an opinion poll, to a participant in a TA project providing input in terms of which questions need to be investigated in the course of a project (e.g., a scoping workshop at the beginning of a TA process).

In the committee model, participation can take the form of a classical hearing. Experts and stakeholders are invited to contribute to the decision-making process by answering the questions of parliamentarians. In general, participatory consultations have become more prominent in parliamentary deliberations. In Germany, for example, the thirty-five-year history of the parliamentary commission of inquiry (*enquete* commission) on S&T subjects can be read as a continuous process of discussing the problem of the public voice in S&T policy making – first as a problem (criticism of S&T) that has to be addressed, later as a voice that has to be heard, and finally as an actor who has to be involved in the work of the inquiry commission itself

(Hennen, Petermann, and Scherz 2004). Another case in point is the instrument of public consultation, more or less standard in the UK Parliament.

In the interactive model of TA, participation still remains part of policy analysis and advice and is not a substitute for decision making. Participation, however, is more than an instrument employed by TA and rather a means to make TA itself public. Meetings such as consensus conferences are organized as publicly visible events to inform policy making directly and stimulate and enlighten public debate about the S&T issue at stake. In this way, participatory TA becomes associated with the concept of deliberative democracy. It is then promoted as a means not only for informing decision making but also for providing both "input-legitimacy" (including a broader range of societal perspectives) and "output-legitimacy" (in the form of heightened legitimacy of policy making through improved decision making). The involvement of social groups in the form of interviews, workshops, and so on has always been part of the TA process. Nevertheless, in the "interactive" or "public" model of TA, society plays a more active role, and participatory methods (e.g., scenario workshops, citizen conferences, public forums) have been systematically developed and applied to give the public a voice and at the same time initiate and stimulate public debate about the issues at stake. For the latter purpose, for instance, science festivals and TV formats have recently been attached to TA processes in the Netherlands.

Is a New Paradigm of S&T Governance Emerging in Europe?

The search for new forms of governance in the field of S&T is ongoing, with particular features, in Europe. These features include a redefinition of the role of (scientific) knowledge and experts in policy making ("democratizing expertise") as well as that of the citizen or general public. This trend toward giving the public a say in S&T has been addressed with the notion of "technological citizenship" (Frankenfeld 1992). This notion implies that the role of the citizen comprises not only civil, political, and social rights but also rights with regard to the development of S&T. Technological citizenship is related to the tendency to regard as politically relevant certain aspects of life that were formerly seen as non-political. The development, diffusion, and implementation of technologies are increasingly being regarded as political issues due to their immense impacts on society. Laypeople are affected by S&T not only as clients (of experts) or consumers but also as members of a polity (citizens). It is thus plausible to regard the indications of a participatory turn that can be observed in S&T policy making in many countries as being technological citizenship in the making.

The ways that citizens are becoming involved differ from country to country depending on its cultural and institutional structures. As Sheila Jasanoff (2005, 249) has shown for biotechnology, there are different ways "in which publics assess claims by, on behalf of, or grounded in science." These ways of making knowledge "socially robust" depend on institutional settings as well as cultural expectations and standards. They include the public role of experts, public accountability, and the transparency of policy making as well as forms of representation that together form civic epistemologies. Such epistemologies – it can be argued – define technological citizenship, which means that they define the position, role, and rights of citizens in the process of societal adaptation of scientific knowledge.

A redefinition of the role of citizens is observable not only at a national level. In the past ten to fifteen years, a series of documents and actions at the European Community level marked a remarkable shift from the previously dominant traditional public understanding of science "deficit model" to a new appreciation of the citizen and his or her views on ethical problems and the risks related to new technologies. There are indications that the predominant technology-driven approach to S&T policy, which includes an instrumental model of technology assessment, has been enriched by efforts to steer S&T in a new direction by making societal needs and demands a part of research agendas. A point in case is the call for dialogue, participation, and empowerment of the European citizen in the EU white paper on governance (European Commission 2001b). Starting from the observation "that people increasingly distrust institutions and politics," the white paper suggests "open[ing] up policy making" to render it more inclusive and accountable (2, 4). The relationship between science and society is regarded as crucial in this respect. A report by the white paper working group, *Democratizing Expertise and Establishing Scientific Reference Systems* (European Commission 2001a), contains the following recommendations: revise the selection of expertise used in the process of policy making, establish guidelines for the selection of expertise, and provide for inclusion of a spectrum of expertise in policy advice that is as broad as possible. Most prominent among the recommendations regarding socially robust knowledge for decision making is the creation of opportunities "for informed participation by society in policy making." The promotion of participatory procedures (e.g., citizen juries and consensus conferences) is one of the means to be employed to support "public debate, knowledge sharing and scrutiny of policy makers and experts" (ii). The European Commission (2001c) took up this reorientation in S&T governance in its *Science and Society* action plan, part of the EC efforts to establish the European Research Area (ERA). The action plan recommends involving people actively in technological development, "particularly in defining the priorities of publicly funded research" (8). To this end, participatory policy making would have "to be widened and deepened to systematically include other sectors of civil society at all stages" (14).

These indications that S&T policy in the European Commission is being opened toward the public have to be viewed in the context of the overall economic objectives that form the guiding perspective of the European Commission's S&T policy and the ERA program. As Levidow and Marris (2001, 345) have argued, "the rhetoric of openness" indicates not a shift to a new contract of science and society but a shift from conceiving controversies over technology as being grounded in the ignorance of the public to a problem of trust in institutions (see also Abels 2002). The shift is thus a way to re-establish trust in policy making by communicating but not by giving up the expert-dominated system of advice. It is really meant not to lead to a reconsideration of the goals and guiding principles of innovation policy but to a means "to restore the legitimacy of science and technology" (Levidow and Marris 2001, 348).

The recommendations of the working group mentioned above, however, show that the European Commission is in need of a reorientation to react to criticism regarding the democratic legitimization of EU policy making, in particular S&T policy. Considering the somewhat contradictory policy objectives (e.g., economic growth versus environmental protection and social integration; forced innovation for global competitiveness versus precautionary principle) and the diversity of policy networks in the European Commission, it is remarkable that the newest insights from the social sciences on the relationship between science and society have had significant influence on the preparation of the white paper.

Participatory TA in Europe: Experience after Two Decades of Experimenting

Given the increased frequency and the widespread use of participatory procedures in TA, one can say that participatory TA in Europe is a success story. A report on participatory TA delivered to the German Parliament that was published in 2004 (Hennen, Petermann, and Scherz 2004) identified seventy-five documented participatory TA projects that were carried out in Europe since the late 1980s. Many more have meanwhile been set up. Furthermore, the first European-wide PTA process (on neurosciences), comprising citizens' conferences in several European countries plus a multi-lingual meeting of 150 citizens from all over Europe, was recently successfully completed (King Baudouin Foundation 2007).

Up to now, however, there has been little systematic empirical evaluation of the role and impact of PTA processes. There are a few documented evaluations of single PTA exercises involving laypersons as citizens (e.g., Enderlin-Cavigelli and Schild 1998; Mayer, de Vries, and Geurts 1995; Mørkrid 2001; Zimmer 2002). These reports clearly show that laypeople can reasonably discuss highly complex societal and ethical aspects of science and technology, that they can enter into a dialogue with experts, and that formats such as consensus conferences are suitable for initiating cooperative learning processes among laypeople. There are thus indications that deliberation – in the sense of joint reasoning on societal problems – can be achieved in PTA processes.

As Rayner (2003) rightly states, however, evaluation has so far mainly concerned aspects of process rather than outcomes or impacts of PTA. Those rare studies that try to assess the political role and function of PTA – such as the EUROPTA project (Joss and Bellucci 2002), based on sixteen case studies of PTA procedures carried out in Europe – indicate that the resonance of PTA in the media and in policy making is often restricted if not almost invisible. Analysis of the factors conducive or obstructive to the public and political resonance of PTA arrangements (Hennen 2002) suggests that the potential influence of PTA is affected by the quality of the outcome of a PTA process or by features of the procedure itself (management, actors involved). More important than features of the procedure, however, is the context in which the procedure takes place. The nature of the issue or problem at stake and the institutional and political setting of the PTA arrangement appear to be of the greatest importance.

The chances for a PTA arrangement to be visible in public or political debates are relatively good when the political situation is open, with relevant actors looking for new ways to solve problems and with no immediate decisions at stake. Typically, these are situations in which the problem has not yet been well defined politically or the relevant actors are searching for common paradigms to solve the problem. The results of PTA arrangements (which include all the relevant stakeholders in the process) have a good chance of being referred to in the public sphere and policy making if the focus is on the development of ideas and the objectives are not highly contested (e.g., the conceptualization of a sustainable city). PTA can have a good opportunity to initialize the learning processes of searching for paradigms and visions (*Leitbilder*) that can be shared by the relevant actors as a common point of reference for further debate. When the issue at stake is highly contested and interest groups hold definite positions with regard to the issue – the more frequent situation in S&T policy making – public debate of the chances and risks might still be going on, but it is highly unlikely that independent policy advice or consultation will have any kind of impact, regardless of whether the advice is expert or participatory.

Regarding the *institutional setting* of a PTA arrangement, the commitment of decision makers to the PTA procedure, the standing and mission of the institution organizing the PTA, and the establishment of PTA as a well-known practice within the country are decisive for public or political resonance. The success of PTA in Denmark is obviously due to the facts that public involvement is a long-standing aspect of Denmark's political culture, that the mission of the institution (the Danish Board of Technology) to organize public debate and provide input to the policy-making process is clear and well known, and that the awareness of PTA in the public and by the political system is high compared with other countries.[1] There are many cases, however, that reveal a lack of commitment to the PTA arrangement by the political system and an unclear mission of the institution, both setting major restrictions on potential resonance in public debate or policy making.

Problems and Challenges: Future Perspectives

The unclear political status of PTA and the fact that it often appears to have little influence on decision making have meanwhile led to criticism. Whereas during the 1990s, normative statements promoting the deliberative quality of participation and its potential to improve the legitimization of policy making were predominant in the literature, more skeptical reasoning and calls for the democratic potential of PTA to be evaluated have recently been heard (Abels 2007; Bora and Hausendorf 2006; Rayner 2003; Stirling 2007). The criticism refers to the facts that:

- the situation of the most important issues at stake in S&T – that is, those contested in society – is not one of openness or a joint search for solutions to problems;
- the commitment of policy making is missing; even worse, PTA might be used only in terms of symbolic politics, allowing for business as usual;
- PTA processes are not open with regard to their outcomes but are meant to re-establish trust in institutions by imposing scientific and risk discourse on laypeople and thus delegitimizing their values and worldviews; and

• the institutional or constitutional role of PTA in established processes of decision making in the representative democratic institutions is unclear or too weak and informal to have any effect.

Although such criticisms are justified with regard to many single PTA exercises, they appear to draw unduly strong conclusions from the observation of the limited effect of PTA. The limited effect is indeed due to the lack of a defined role for PTA in the established decision-making processes and not caused by an inborn bias of PTA procedures. It can be shown that the outcomes of many PTA procedures (i.e., their conclusions or recommendations) often contradict the expectations held by experts and decision makers. The argument that PTA is a means of framing issues in such a way that non-scientific arguments (considered to be irrational) are ruled out by the process, or that citizens are involved in a procedure that rules out their authentic attitudes, does not find much support in empirical studies of the outcomes of citizen juries in contested fields of technology or scientific development. Dryzek et al. (2009), in their exploration of mini-publics (citizen juries, consensus conferences) on GM food in seven countries, found for example that precautionary worldviews are pervasive in deliberating publics – which means that laypeople prefer to be cautious with regard to uncertainties about possible harm and are at least skeptical of the marketing of GM food. They strongly oppose Promethean worldviews, which are based in "faith in the capacity of humans to manipulate complex systems" (206) popular among political elites and GM food promoters. Dryzek and colleagues conclude from their analysis that "Promethean aspirations" for GM food cannot be legitimized by mini-publics (284).

Framing and the instrumental use of outcomes are, as Stirling (2007) has argued, a problem for any type of policy advice and technology appraisal. He holds that participatory appraisal is as open to power and justification strategies as is expert appraisal. PTA can be used to induce "technical commitments": that is, closing processes of technology development instead of opening them up to new perspectives and values. Whereas expert appraisal can be framed by strategies such as prioritizing research, accrediting expertise, and recruiting committees, the same can be achieved for participatory appraisal by strategies such as structuring process design, recruiting participants, and phrasing questions. Stirling is right in pointing at the vulnerabilities of deliberate appraisal. Most PTA practitioners would agree with him when he states that "it seems clear that the apparent normative democratic credentials of participatory appraisal do not themselves confer immunity to instrumental pressures for the justification of powerful interest" (277). How-ever, it is by providing for transparency, and by sharing control of the process with participants, that practitioners strive for an open and non-biased pro-cess. In brief, there are rules of good practice (some of them derived from Habermasian discourse ethics) that can protect against instrumentalization and framing. With regard to expert appraisal (science), one cannot deny the relevance of central institutional features of science such as peer review and methodological skepticism despite known cases of scientific fraud. Similarly, participatory appraisal is guided by rules of discourse and the principle of transparency, functional equivalents to the Mertonian principles of science.

Critical reasoning makes us aware of the vulnerability of PTA and has to be taken seriously. Recent criticism, however, partly ignores the fact that PTA, as an element of deliberative democracy, *necessarily* has to act in an environment dominated by political cultures, institutions, and powerful actors who are often hostile to any restructuring of science and research policy making. We have to acknowledge that PTA is part of the real world and makes up *only one aspect* of an ongoing movement toward more demo-cratic structures in S&T. PTA is a new form of governance but not of polit-ical mobilization, as Rayner (2003) apparently would like to see it. It has emerged in reaction to a crisis of representative structures of policy making but has not yet found and defined its role. In the remainder of the chapter, I will discuss two problems of PTA that are related to technological citizen-ship and that, I think, have to be dealt with in the future to sharpen PTA's political profile and role.

Civic Epistemologies: Clarifying the Role of Citizens

Given the cognitive uncertainties and normative ambiguities associated with S&T, an appraisal of its consequences that goes beyond the compe-tence of experts is needed and must include the public in general. This is, as it were, the standard argument for participation in S&T policy making. What the new quality is in terms of the knowledge that laypeople actually contribute, however, is not sufficiently clear in practice, and research has not sufficiently studied it.

We need to clarify what civic epistemologies actually include to better understand what we can expect from different forms of a participatory ap-praisal of technology. A first, instructive step toward this aim has been undertaken by Evans and Plows (2007), based on a typology of different levels of expert knowledge suggested by Collins and Evans (2002). They argue – drawing on their experience with PTA processes in the United

Kingdom – that referring to the "public" as opposed to "experts" (as is often the case in PTA discourse) misses the different types of knowledge that must be taken into account – from that of technical experts, to that of activists who have to be regarded as lay experts, to that of laypeople. A true layperson is characterized by not disposing of specific knowledge, only of ubiquitous or popular knowledge. The disinterestedness or impartiality of laypeople (their "being not involved") is regarded by Evans and Plows as the central feature that enables laypeople to develop an external "meta perspective" when dealing with contested issues of science and technology. When properly informed, they can appraise expert knowledge with regard to its practical (everyday life) implications. The appraisal is based either on general social beliefs (ubiquitous discrimination) or on particular knowledge and beliefs that they hold as members of a particular community (local discrimination). The lack of a specific commitment enables laypeople to act as an independent jury, hearing evidence from different experts and knowledge cultures.[2] "Only those who are situated outside the committed knowledge cultures of both the scientific and activist communities can bring in a genuine civic epistemology" (Evans and Plows 2007, 843).

The plea of Evans and Plows for fostering the role of the citizen indicates at least the direction that further exploration of the contents and structures of civic epistemologies should take. A civic epistemology, like any, comprises not only manifest inventories of knowledge but also methods and beliefs separating the relevant from the irrelevant, the true from the false, and so on. For a proper analysis, then, reference to concepts of everyday knowledge and orientation is needed (making use of theories of everyday life as provided by, e.g., Alfred Schütz [1946] or ethnomethodology) as well as a differentiated analysis of civic epistemologies in relation to different types of problems and issues dealt with in PTA.

The preliminary thoughts given in Table 2.1 on the different roles of citizens or the types of citizenship with regard to the involvement of laypeople in S&T are meant to sketch the scope of analysis needed. If we look at types of lay involvement in S&T, we can discern the consumer, the stakeholder or local citizen, and the well-informed citizen. As consumers, laypeople are related to S&T issues by their personal interests as actors in a market economy. They bring in no other knowledge than that based on common life experiences and beliefs. When involved in participatory procedures – such as a public consultation or survey on GM food – they are asked to (and are competent to) express their individual preferences.

TABLE 2.1

Aspects of technological citizenship

Knowledge			Interests	Values/ ideas	Political status (citizenship)
Ubiquitous	Local	Meta			
Consumer					
Experiences and beliefs			Personal	Individual preferences	Bourgeois (market)
Local citizen/stakeholder					
Experiences and beliefs	Specific group knowledge		Personal and local	Local "we"	Holder of rights (state of law)
Well-informed citizen					
Experiences and beliefs		Informed common sense	Public interest	Ethics of "good life" community	Member of political

As members of an affected group of citizens, as defined either by local residence or by shared interest of other origin (e.g., patients), laypeople can give specific knowledge input based on their local experiences (the Sellafield example; see Wynne 1996). They share interests (as residents affected by a planning process), and they *know* that they share interests and thus define themselves as a "common we" who demand the right to make autonomous decisions (Bora and Hausendorf 2006). Their citizen role is defined by the rights that they hold, which can be inscribed in a legal framework (e.g., in planning or environmental law). Being experts on their own housing and living situations, they can play an active role in planning processes. Additionally, their bargaining power can be relatively high with regard to public authorities and administration. As directly affected residents, laypeople can block technical solutions that are not convenient to their interests by initiating joint public actions or going to court. What ideally can be achieved is a "public inquiry" (Fischer 1999) type of PTA: cooperation between lay experts and scientific experts in searching for ways to optimize the use of expertise for social purposes.

As well-informed citizens, laypeople have no particular knowledge to contribute, but, as disinterested and informed citizens, they can have a

metaperspective and appraise S&T from a common-sense perspective, independent of any particular knowledge community. This reflects their personal preferences in the light of public interests and an ethical consideration
of the issues at stake with regard to the concept of a good life. Thus, in PTA,
they can act as members of a political community that has to come to terms
with the challenges and uncertainties induced by advanced S&T. The presence of this perspective cannot be taken as a given but has to be assured by
a discursive arrangement. The well-informed citizen is not asked to dismiss
his or her own personal interests and ideas but to reflect on them in the light
of other cognitive and normative claims.

PTA and Decision-Making Processes: Formal or Informal?

PTA is obviously not a substitute for representative decision-making procedures, but it is meant to inform and supplement decision making in representative democracies. Participation in the context of TA is more a form
of inclusive policy analysis than a means of decision making. PTA has nevertheless suffered because it is often hard to identify how it makes a difference:
that is, what resonance it has produced in the established policy-making
structures. It often seems to be an idle exercise, an insignificant addition to
established decision-making routines. No matter how democratically, fairly, and transparently the process has been organized, it can end up as did a
Danish conference on GM food: "To the extent that the conference was
brought into political communication, it was mostly as a means to claim
that due attention had been paid to the public's concerns, not as an occasion
to modify policies or regulatory principles" (Hansen 2006, 578).

An evaluation of the effects of PTA must take into account that, as is well
known from knowledge utilization research, it is difficult and often impossible to identify the effects of scientific advice on decision-making processes. Such processes are generally influenced by a complex set of interests and
rationalities (Albaek 1995). Thus, expert appraisal suffers from the same
lack of a visible impact as does participatory appraisal. In contrast to expert
appraisal, however, PTA cannot simply be regarded as a (paid) service to
policy making (that might be listened to or not). PTA bears, by including
underrepresented societal perspectives, a connotation of democratizing
S&T policy. It produces expectations of democratic inclusion. The lack of a
perceptible impact is therefore more critical to PTA than to expert appraisal
and might, in the long run, create grave disappointments.

This is a serious challenge to the future of PTA. Either PTA redraws its
aspirations drastically, becoming just a sophisticated means of policy-

related social research, or it has to find a way to attach itself in a more formal way to established representative institutions and decision-making processes.

I want to briefly outline the fundamental problem of defining "techno-logical citizenship" connected to this question. I will do so by drawing on the findings of an in-depth conversation analysis of participatory procedures on field releases of genetically modified organisms (GMOs) in seven countries (Bora and Hausendorf 2006). The participatory procedures were part of for-mal licensing procedures embedded in a legal framework, which means that participatory meetings were part of an official formal decision-making pro-cedure. Bora and Hausendorf show that different types of participants held different concepts of technological citizenship and how, in the course of communication, the concept of technological citizenship was constructed by the procedure. "Citizenship" was thus understood "as a semantics that has to do with the inclusion of persons in the political system" (480). Deci-sive for this inclusion are the concepts of citizenship held by different actors in public debates and inscribed in the administrative procedures and legal frameworks of which the participatory procedure is a part. The striking re-sult of the analysis is that the formal procedure that is part of the legal frame-work and focuses on risk analysis clashes with the concept of technological citizenship held by laypeople. Whereas citizens expect the procedure to in-clude negotiations on aspects such as values, power, and justice, the proced-ure itself and the main actors systematically exclude value-laden arguments and try to restrict the process to what Bora and Hausendorf (2006, 485) call "questions of truth" – scientific questions of risk assessment. I cannot go into the details of the analysis, and I factor out the authors' questionable separation of truth from values, which I find problematic. Relevant in the present context is that Bora and Hausendorf show that the formal *role* and position given to citizens in the licensing procedure bind them to the rules of the procedure. These rules, together with the related concepts of citizen-ship held by actors such as the leading authority, systematically rule out the citizens' *arguments* (the "contents" of citizenship, as it were) as illegitimate. The authors conclude that the type of technological citizenship constructed in such formal participatory procedures creates a dynamics of exclusion and systematically produces disappointment among laypeople. For this reason, it is dysfunctional with regard to the legitimization of decision making. I agree with their view, but I have difficulties following their (implicit) conclusion, namely, that PTA should restrict itself to an informal role to be more inclu-sive of laypeople's needs and demands and not produce disappointment.

TABLE 2.2

The participation dilemma

Type of participatory procedure	Two aspects of citizenship	
	Role inclusion	Discourse inclusion
Formal/legal procedure	High	Low
Informal procedure	Low	High

What we learn from the many cases of more informal PTA procedures is that they are in danger of becoming politically meaningless and without consequence. If we distinguish between the role and the discourse aspects of technological citizenship, the latter including value-based arguments, then we see what one might call a dilemma of participation (see Table 2.2). In formalized participatory procedures (those with some legal or constitutional status within the established system of decision making), high inclusiveness regarding role comes at the price of the exclusion of central arguments. In contrast, informal procedures are inclusive (or at least can be) with regard to arguments, values, and perspectives, but they are often detached from decision making and, thus, are exclusive of a formal role or position in decision making.

We have to face the fact that this dilemma can only be overcome on the basis of a high degree of public awareness, vivid public debates, and a broad discourse on governance within the representative system. A process of social and cultural change might lead or force decision makers to seriously rethink their concepts of technological citizenship and become open to institutional arrangements that position PTA clearly within decision-making structures.

NOTES

1 Mejlgaard (2009) has shown that in Denmark, changes in the political context and framing are possible and might lead to a revitalization of an existing undercurrent of standard public understanding of science.

2 The founder of phenomenological sociology, Alfred Schütz, came to similar conclusions when differentiating the well-informed citizen (to whom it falls to determine which experts are competent) from the (uninterested) "man in the street" and the specialized expert in his work "The Well Informed Citizen: An Essay on the Social Distribution of Knowledge," originally published in 1946.

REFERENCES

Abels, G. 2002. "Experts, Citizens, and Eurocrats: Towards a Policy Shift in the Governance of Biopolitics in the EU." *European Integration Online Papers* 6, 19: 1-26. http://eiop.or.at/eiop/.

–. 2007. "Citizen Involvement in Policy Making: Does It Improve Democratic Legitimacy and Accountability? The Case of PTA." *Interdisciplinary Information Sciences* 13, 1: 103-7.

Albaek, E. 1995. "Between Knowledge and Power: Utilisation of Science in Public Policy Making." *Policy Sciences* 28: 79-100.

Beck, U., A. Giddens, and S. Lash. 1994. *Reflexive Modernization: Politics, Tradition, and Aesthetics in the Modern Social Order.* Cambridge, UK: Polity Press.

Bora, A., and H. Hausendorf. 2006. "Participatory Science Governance Revisited: Normative Expectations versus Empirical Evidence." *Science and Public Policy* 33, 7: 478-88.

Collins, H.M., and R. Evans. 2002. "The Third Wave of Science Studies: Studies of Expertise and Experience." *Social Studies of Science* 32: 235-96.

Dryzek, J.S., R.E. Goodin, A. Tucker, and B. Reber. 2009. "Promethean Elites Encounter Precautionary Publics: The Case of GM-Foods." *Science, Technology, and Human Values* 34: 263-88.

Enderlin-Cavigelli, R., and P. Schild. 1989. *Publiforum Strom und Gesellschaft: Evaluationsbericht der Stiftung Risiko-Dialog.* Schweizer Wissenschaftsrat, Document de Travail TA-DT 21.

European Commission. 2001a. *Democratizing Expertise and Establishing Scientific Reference Systems.* Report of the Working Group on the White Paper of Governance. Brussels: EC.

–. 2001b. *European Governance: A White Paper.* Brussels: EC.

–. 2001c. *Science and Society: Action Plan.* Brussels: EC.

Evans, R., and A. Plows. 2007. "Listening without Prejudice? Re-Discovering the Value of the Disinterested Citizen." *Social Studies of Science* 37: 827-54.

Fischer, F. 1999. "Technological Deliberation in a Democratic Society: The Case of Participatory Inquiry." *Science and Public Policy* 26: 294-302.

Frankenfeld, P.J. 1992. "Technological Citizenship: A Normative Framework for Risk Studies." *Science, Technology, and Human Values* 17: 459-84.

Guston, D.H., and B. Bimber. 2000. *Technology Assessment for the New Century.* New Brunswick, NJ: School of Planning and Public Policy, Rutgers University.

Hansen, J. 2006. "Operationalising the Public in Participatory Technology Assessment: A Framework for Comparison Applied to Three Cases." *Science and Public Policy* 33, 8: 571-84.

Hennen, L. 1999. "Participatory Technology Assessment: A Response to Technical Modernity?" *Science and Public Policy* 26: 303-12.

–. 2002. "Impacts of Participatory Technology Assessment on Its Social Environment." In *Participatory Technology Assessment: European Perspectives,* edited by S. Joss and S. Bellucci, 257-75. London: University of Westminster Press.

Hennen, L., T. Petermann, and C. Scherz. 2004. *Partizipative Verfahren der Technikfolgenabschätzung und Parlamentarische Politikberatung.* Office of

Technology Assessment at the German Parliament, Working Report 96. English summary at http://www.tab.fzk.de/.

Jasanoff, S. 2005. *Designs on Nature: Science and Democracy in Europe and the United States.* Princeton and Oxford: Princeton University Press.

Joss, S., and S. Bellucci, eds. 2002. *Participatory Technology Assessment: European Perspectives.* London: University of Westminster Press.

King Baudouin Foundation. 2007. *Meeting of Minds.* http://www.kbs-frb.be/.

Levidow, L., and C. Marris. 2001. "Science and Governance in Europe: Lessons from the Case of Agricultural Biotechnology." *Science and Public Policy* 28: 345-60.

Mayer, I., J. de Vries, and J. Geurts. 1995. "An Evaluation of the Effects of Participation in a Consensus Conference." In *Public Participation in Science: The Role of Consensus Conferences in Europe,* edited by S. Joss and J. Durant, 109-25. London: Science Museum.

Mejlgaard, N. 2009. "The Trajectory of Scientific Citizenship in Denmark: Changing Balances between Public Competence and Public Participation." *Science and Public Policy* 36: 483-96.

Mørkrid, A.J. 2001. "Consensus Conferences on Genetically Modified Food in Norway." In *Citizens as Partners: Information, Consultation, and Public Participation in Policy Making,* 223-37. Paris: OECD.

Nowotny, H., P. Scott, and M. Gibbons. 2001. *Rethinking Science: Knowledge and the Public in an Age of Uncertainty.* Cambridge, UK: Polity Press.

Rayner, S. 2003. "Democracy in the Age of Assessment: Reflections on the Roles of Expertise and Democracy in Public-Sector Decision Making." *Science and Public Policy* 30: 163-70.

Schütz, A. 1946. "The Well Informed Citizen: An Essay on the Social Distribution of Knowledge." *Social Research* 13: 463-78.

Stirling, A. 2007. "'Opening Up' and 'Closing Down': Power, Participation, and Pluralism in the Social Appraisal of Technology." *Science, Technology, and Human Values* 33, 2: 262-94.

Vig, N., and H. Paschen, eds. 2000. *Parliaments and Technology: The Development of Technology Assessment in Europe.* New York: State University of New York Press.

Wynne, B. 1996. "Misunderstood Misunderstandings: Social Identities and Public Uptake of Science." In *Misunderstanding Science: The Public Reconstruction of Science and Technology,* edited by A. Irwin and B. Wynne, 19-46. Cambridge, UK: Cambridge University Press.

Zimmer, R. 2002. *Begleitende Evaluation der Bürgerkonferenz "Streitfall Gendiagnostik."* Fraunhofer Institut Systemtechnik und Innovationsforschung, Karlsruhe.

Democracy, Governance, and Public Engagement
A Critical Assessment

PETER W.B. PHILLIPS

45-65

The number of countries attempting to run their governments on democratic principles has been on the rise over the past thirty years (Polity IV), but the performance of those governments has not met the rising expectations of individuals and groups in society. This is especially a concern in many OECD countries as new, transformative technologies are proposed by industry and government as solutions to long-standing problems or as golden opportunities to break loose from the constraints of existing production possibilities. Ultimately, individuals and groups want the governing system and its policies and decisions to reflect their values, interests, and beliefs. The rising complexity of the issues and the increasing diffusion of perspectives have swamped most governments. In response, a wide range of engagement models has been developed and tested in a selection of policy areas in an attempt to bridge the gap between the ideal and the practical.

This chapter reviews the scale of the challenge, examines the underlying concepts and principles of democracy, and unpacks the critiques of government that have led to the rise of democratic engagement as a new form of governing. The chapter then investigates a range of evaluative frameworks that one might use to assess the appropriate use and contributions of various engagement methods and applies a number of approaches to the main democratic engagement options used for biotechnology policy. The chapter concludes with an assessment of what further might be needed to fully understand the appropriate role and function of public engagement in democratic systems.

The Convergence of Democracy and Democratic Engagement

In one sense, democracy and democratic engagement are both very old and very new. From the beginning in ancient Athens (fifth century BCE), democracy – "rule by the ruled" (Brown 2009) – was inextricably dependent on a highly aware, capable, and engaged group of citizens. In the interregnum, as democracy fell into disuse and monarchies (rule by one) and oligarchies (rule by a few) became the dominant system, the resulting hierarchies and other power relationships did not require extensive public engagement in the governing system. Elites led, and the rest followed.

With the Enlightenment and the articulation of a liberal, humanist conception of government and the emergence of the Westphalian system of states (via the Treaty of Westphalia, 1648), the monarchies and empires of the Middle Ages eventually gave way to nation-states that increasingly demanded new forms of government to reflect their new status. Democracy returned. The earliest modern forms of democracy, as embodied in the United Kingdom, the United States, and France, were quite different from what we see and expect today. It was only in the past century that women, certain minorities, and young adults were offered the franchise in these countries. Now, universal suffrage and the right (if not expectation) that individuals may individually or collectively organize to seek and use the levers of power are two main elements that one expects in any democracy.

The Polity Project has a long-standing research tradition of coding the authority characteristics of states in the world.[1] The current version of the project has collected data on all major, independent states in the global system (i.e., the 163 states with a total population of at least 500,000) over the period 1800-2008. The Polity scheme scores national governments based on six components that focus on the qualities of executive recruitment, constraints on executive authority, and political competition. The "Polity Score" captures this spectrum on a twenty-one-point scale, ranging from –10 (hereditary monarchy) to +10 (consolidated democracy). Democracies score +6 to +10 on their scale. The Polity IV dataset reports that ninety-five countries are democratic, up from approximately forty in 1980. Over the corresponding period, the number of autocracies has declined from more than ninety to fewer than thirty. Although none of the continents is universally democratic, the number of democracies is rising in all regions, and democracy is in the ascendancy and the leading form of government in all continents except Africa, where monarchies and autocracies remain resilient.

As issues have become more complex and citizens have become more educated and aware of the issues facing governments, these democracies have been challenged to become more open, responsive, and reflexive. Both corporatist and pluralist/interest group models have been tried, but many citizens have been unable to access those highly structured subsystems and thus remain disenfranchised. In response, governments in the past generation or so have sought new ways to engage with citizens on issues that they care about. A quick Google search in late 2009 revealed more than 8 million pages related to public consultation and more than 7 million related to democratic engagement.

There have been a number of collations of the most popular types of consultation and engagement methods. Most of the categorizations simply look at the scale, scope, and purposes of the exercises. Rowe and Frewer (2005) offer a more nuanced typology based on flows of information – one-way flows from sponsors to citizens (i.e., communication) or from citizens to sponsors (i.e., consultation) and two-way flows between the sponsor and citizens (i.e., participation) – that are then differentiated based on whether the information to be exchanged is fixed or flexible, whether the structure of the process and the participation are open or closed, and whether the aggregation of opinions is structured or flexible. The authors identify four types of communication processes, six types of consultation processes, and four types of participation process. With the advent of new media and communication systems, mechanisms are as varied and numerous as the issue, audience, and sponsor, ranging from print, audio, and visual to multi-media offerings, with real-time interactive processes to deal with consultations.

Although a typology and a listing of the specific types of mechanisms are valuable, in a way, the analytical approach is inverted, tending to assume that the answer is some form of consultation or public engagement process and that the only remaining issue is to select one and figure out how to use it most efficiently and effectively. What seems to be missing is a more fundamental and in-depth discussion of the problem that democratic engagement is intended to fix. In short, if democratic engagement is the answer, then what is the question?

Democracy and Its Limits

Phillips (2007) asserts that governments are fundamentally faced with a complex systems problem when considering how to govern transformative

technological change. The institutional challenge of governing transformative innovation requires the involvement of three differently conceived domains. Economist Kenneth Boulding (1970) characterizes these domains as
the compulsory, the contractual, and the familistic. The compulsory system
involves threats (e.g., "you do something for me, or I'll do something nasty
to you") that fundamentally depend on the credibility of the threat and the
capability of the partners to affect their sides of the relationship. These types
of arrangement can yield either zero-sum or negative-sum results. The contractual, exchange-based system involves voluntary bids, offers, and transactions, which generally yield exchanges that benefit both the buyer and the
seller. The familistic, integrative system involves formulating communal
"voice" through status relationships (e.g., "you do something because I am
your father, or a king, a priest, a teacher, a lover, a child, a student ... "). This
yields three different methods of integration: coercive relations that distribute rights and obligations, led by the state; *quid pro quo* exchanges in the
market governed by Marshallian supply and demand; and voluntary dealings, in which cooperation, reciprocity, and solidarity engage community
and society (Paquet 2001). Boulding (1970) argues that society can be
viewed as a triangle (his "social triangle") in which all organizations – including the state, the market, and civil authorities – are built on one or a
balance of the three relationship systems.

Governing innovation thus ultimately involves both economic and non-
economic actors and incentives. Who leads will depend on the balance of
three factors: rivalry (does the activity create congestion costs?); excludability (can people be kept from using and benefiting from an activity?); and
voice (does the activity require specialized, intermediate, contextual knowledge from others?). Depending on the dominance or balance of these factors, public state, private market, or collective authorities might offer the
most effective loci for governance.

The triad of institutions – governments or states, the market, and social
or familial organizations – have specific institutional attributes that make
them more effective at producing particular types of goods (Picciotto 1995).
The government sector is best at producing public goods – low excludability
impedes private markets, while the low voice component makes it difficult
for the collective sector to organize. The private sector tends to dominate
whenever property rights can be assigned to make rival goods excludable,
thus enabling firms to sell at the marginal cost of production. The participatory sector is best at governing common pool goods (e.g., standards and

norms); a collective will usually have enough information to enable it to effectively manage the resource and capture the benefits.

In a perfect world in which all activities would be parsed into one of the three pure domains – as public goods, private goods, and collective or common pool goods – there would be no need for public engagement. Governments would be efficiently and effectively elected by universal suffrage and would offer optimal amounts of public goods. Markets would produce, and consumers would buy optimally, with individuals making choices based on their own preferences, full information, and prices that would fully reflect all the costs and benefits of those choices. Groups would form and offer the optimal range of common pool goods and services, for there would be no free riders. But the world is far from perfect. States, markets, and collective authorities all exhibit flaws that can necessitate democratic engagement.

States are challenged to meet the democratic ideal. Political scientist Robert Dahl (1998) proposes five standards for democracy: equal and effective opportunities for participation; equal voting; equal and effective opportunities for learning about policy options and implications; membership control of the agenda; and universal suffrage. To deliver these standards, a modern, large-scale democracy would then need six key political institutions: elected officials; free, fair, and frequent elections; freedom of expression; alternative sources of information; free right of association; and inclusive citizenship. There is a wide range of explanations of exactly how such a system would work and whether it could actually sustain participation by citizens and deliver outcomes that improve the greater good. There are two main critiques of this model. Kenneth Arrow (1951), using pure public choice theory, mathematically demonstrated that a democratic system is over-specified and bound to fail. He posited a case in which a number of individuals can each rank his or her preferences on a given set of outcomes. Based on a strict set of criteria for a "fair voting method" (mainly that all preferences of all voters in the ranking/voting are included, that there is universal suffrage, and that strict *Pareto* criteria of doing no harm must be met), it is impossible to design a social welfare function that will satisfy all the criteria at once. Translated into voting, it simply means that a majority voting system will not deliver optimal social welfare – the majority will vote in such a way as to reduce the net value of the outcomes. Amartya Sen (1979) relaxed some of the assumptions to yield a slightly better outcome, but his work still showed that majority voting can often lead to perverse results. Anthony Downs (1957) offered one explanation for how this might occur by looking at the voting system itself. He constructed a comprehensive theory of democratic decision making, assuming that both voters and politicians will engage in self-serving behaviour, such that electoral candidates and parties would locate themselves in the political spectrum to maximize their chances of being elected. This median voter model asserts that successful candidates tend to take a position at the median of a normal distribution of voters, which tends to drive the democratic system to deliver median views, which at times can diverge significantly from the national average and is likely to deliver results that are not fully commensurate with the public good. Downs further suggested that prospective voters in such a system will have limited incentives either to inform themselves about the issues or to vote (Hardin 2002).

Perhaps the biggest challenge to modern democracies is that the concept of governing has been recast as governance. British political scientist Rod Rhodes (1995, 1) suggests that governance is actually "a new process of governing; or a changed order condition of ordered rule; or the new method by which society is governed." It is not a synonym for government but involves a new system of "self-organizing networks or 'governing without government.'" Rhodes identifies a number of functional uses of the term "governance" that challenge the democratic state. Governance could entail a minimal state with a new corporate-style government; new public management, involving professional management; a shift from rowing to steering (Osborne and Gaebler 1992); a socio-cybernetic system that is distributed and "centreless" (involving subsidiarity, absence of a single sovereign authority, multiplicity of actors, interdependence, and blurred boundaries); and self-organizing networks, involving interdependent actors that cannot be steered. Increasingly, there is agreement that governance does not rest on the authority and sanctions of government (Stoker 1998, 17). The concept refers instead to a set of institutions and actors drawn from government and beyond, involves blurring of boundaries and responsibilities, entails power dependence between institutions, and encompasses autonomous, self-governing networks of actors and works without the power and authority of government.

Markets present similarly intractable challenges. Even in a libertarian world, there is an economic rationale for the state to engage in the economy and society when there are public good failures (e.g., private initiative would produce too little of a non-rival, non-excludable good, such as public health or national defence), externalities (e.g., either too little of a good is produced [research] or too much is produced [industrial pollution] because the market is unable to incorporate the spillover effects into the market price), or full and equal access to information about a good and its impacts is not forthcoming (e.g., imperfect or asymmetric information). In those cases, the state might subsidize, tax, or regulate activities to improve their production, distribution, and transparency.

Finally, the "third sector" faces real challenges in appropriately mobilizing the public. Many date the formal study of organizations in the political system to Arthur Bentley, who in 1908 wrote a book credited with initiating "group theory" in modern political studies. He urged study of the political process in terms of groups, interests, and pressures, where politics and policy were nothing more than reflections of group competition in the political process. David Truman in 1951 revived Bentley's group process theory of government to examine the rising role of interest groups in the United States and abroad. This approach offered a tool for analysis, a theory to drive systematic behavioural research, and a range of testable hypotheses related to the political orientations of groups, the internal politics of groups, and the influence of groups on the legislative, executive, judicial, and electoral systems. The difficulty with this work is that it assumes that it is both obvious and rational for individuals to act collectively; there is little explicit examination of the motivations and choices with which individuals might be faced. In 1965, economist Mancur Olson produced *The Logic of Collective Action: Public Goods and the Theory of Groups,* which posed a truly behavioural theory of collective action. Assuming that individuals are self-interested actors who will undertake some form of introspective optimization calculus about whether to join a group or not (similar to what is assumed to happen when individuals make consumption decisions), small groups might form spontaneously due to the difficulty to free-ride, but larger groups often fail to form whenever the output of the group exhibits positive externalities and is not excludable. Thus, the theory goes a long way toward explaining the emergence or failure of many interest groups, epistemic communities, and some values-based groups (Phillips 2007). Olson's logic, however, requires rationality in the economic sense, in which individuals undertake some form of selfish optimization calculus. This might not hold true, however, for many philanthropic or religious lobbies or even for many social action mass movements seeking to further their values and beliefs.

All of these failures or limitations suggest that a significant portion of a population in a modern industrial democracy will effectively be disenfranchised from decision making. This is especially problematic with decision making regarding the introduction of new transformative technologies and their products because they are difficult or impossible to fit into the voting structure and exhibit credence attributes and externalities in the market. While there might be value in mobilizing voice, there is also a real potential for free riders. In short, pure archetypes of states, markets, and civil authorities cannot meet the demands of new technologies.

One additional argument has been offered recently. The emergence of the post-positivist, postmodern critique suggests that the issue is bigger than making the other three systems work better. Rather, the critique fundamentally rejects the cult of expertise that underlies the modernist approach laid out above. This view asserts that the key to success is to instill a new level of institutional fairness. In short, democracy, the market, and the collective sectors are deemed to fail unless full participation is realized (Howlett, Ramesh, and Pearl 2009). Although the critique largely rejects the aforementioned modernist approach, the underlying flaws in the state, market, and civil authorities remain.

The rest of this chapter attempts to assess whether public engagement in all its varied forms helps to alleviate the deficits of democracy, the market, and civil authorities – or whether it simply creates a new set of problems and failures.

If Deficits Are the Problem, Is Democratic Engagement the Solution?

Complex governing systems, in particular, are challenged to remain accountable, responsible, and transparent. Nested systems involving complex subroutines are always eyed suspiciously by the average citizen, particularly if the procedures are poorly defined and the decisions are difficult to understand in context. Many have called for our governing systems to become more reflexive. At one level, this makes some sense, for the risks and benefits of any transformative technology are usually unknown, and probably unknowable, by any narrow set of governors. A wide number of actors from the commercialization system, ranging from scholars and researchers in the epistemic community to ultimate consumers and citizens, can and probably should have a chance to contribute to our assessments of transformative change.

The early body of literature and debate about engagement models focused on corporatist or neo-corporatist engagement – formal engagement between unelected corporations or interest groups representing economic, industrial, and professional groups in the legislation and regulation of the state. This approach, drawing on roots laid down in the medieval guilds, was developed and propounded in nineteenth- and twentieth-century Italy and Germany as an alternative to market-based systems or socialism. Generally, corporatist states recognize dependence on corporate actors but also seek to use them as instruments to both effect and legitimate policies. More recently, the literature has identified and decried a creeping corporatism in liberal democracies, through lobbying, regulatory capture, and the sometimes dominant role of the military-industrial complex. Neo-corporatism has emerged more recently as a set of new social arrangements involving tripartite bargaining among unions, the private sector, and government. Various attempts have been made to bring the three parties together nationally (e.g., in many European countries, such as Germany) or internationally to deal with wage bargaining and facilitate major economic transitions. One area that many point to in terms of emerging corporatism is in the management of international trade and the international financial system; recent analyses suggest that, rather than a true corporatist model emerging, we might actually be moving toward a multi-layered system of governing in these areas (Held et al. 1999). Although corporatism and neo-corporatism offer some additional points of access to address the deficits in the state and market, the opportunities for average citizens remain limited – one needs standing in one of the sectors to have a seat at the table.

One response has been for governments to directly measure the interests of individuals. Academics and governments have invested heavily in new economic and social analytical tools, such as cost-benefit, input-output, and revealed preference studies in an attempt to gather evidence of individual interests to support decision making related to policies, programs, and projects. Although these methods do expand the array of public views in the decision-making process, framing public views solely in economic terms often narrows and distorts them.

More recently, governments have worked to nurture the growth of civil authorities that can then engage in the political process. Governments around the world have provided subsidies, preferential tax treatment, contractual support, and privileged access to public debate to encourage individuals to organize, in an attempt to overcome the latent free-rider problem inherent in civil authorities. Although civic organization itself might be valuable, since it creates both bonding and linking social capital that is vital to sustaining the civic society and effective government (Putnam 2000), it is not clear how it contributes to resolving the deficits in the three governing domains in modern industrial democracies.

TABLE 3.1

Assessing engagement as a contribution to democratic norms

	Equal participation	Equal voting	Learning opportunities	Membership control of agenda	Universal suffrage
Referenda	High	High	Low–medium	Low	High
Public hearings	Low–medium	Low–medium	Medium	Medium	Low
Public opinion surveys	High	High	Low	Low	Low
Negotiated rule making	Low	Low	Medium–high	High	Low
Consensus conferences	Low–medium	Low	High	Varies	Varies
Citizens' jury/panel	Low–medium	Medium	High	Varies	Varies
Citizens' advisory committee	Low–medium	Low	High	Varies	Varies
Focus group	Low–medium	Low–medium	High	Varies	Varies
Expert advisory group	Low	Low	High	High	Low

Source: Based on criteria offered by Dahl (1998).

Although the number of participation mechanisms is apparently large, with many different types of meetings, workshops, conferences, and fora, they exhibit many common elements and often are interchangeably named. Rowe and Frewer (2000) suggest that an appropriate array of suitably diverse models of public engagement would involve referenda, public hearings, public opinion surveys, negotiated rule making, consensus conferences, citizen juries or panels, citizen advisory committees, and focus groups. One additional model – the expert advisory group augmented by one or more representatives of the *vox populi* – is mooted by many as a proxy for public input. These nine methods are the focus of the analysis in this section.

There are a number of ways in which one might evaluate public engagement processes, including using benchmarks offered by Dahl (1998), Rowe and Frewer (2005), Fiorino (1990), Pal and Maxwell (2004), and Castle and Culver (2006).

First, one could put public engagement processes up against Dahl's five standards for democracy: equal and effective opportunities for participation, equal voting, equal and effective opportunities for learning about policy options and implications, membership control of the agenda, and universal suffrage. In Table 3.1, the results of a qualitative assessment of the nine key methods of democratic engagement against the democratic norms illustrate that, though some approaches address aspects of the democratic deficit, none unambiguously meets all of Dahl's norms. Generally, there is a trade-off between representativeness and control of the agenda; the more the systems attempt to incorporate and represent the array of opinions in society, the more the structure of the processes is controlled by others, and vice versa. On the face of it, none of the key methods is a sure way to deal with the democratic deficit.

Second, Rowe and Frewer (2005) offer a framework for evaluating public participation techniques using, among others, the concepts of accountability (independence of the participants), responsibility (representativeness of the participants), and transparency (transparency of the process to the public). They then take eight of the possible methods commonly used and rank them against the ARTful criteria (see Table 3.2). Although every method scores moderately or highly on at least one of the foundational criteria, none scores universally high.

Third, Daniel Fiorino (1990) of the US Environmental Protection Agency suggests that participation theory offers four criteria for evaluating democratic engagement processes. First, a mechanism should allow the direct participation of amateurs in decisions. Second, citizens should share in collective

TABLE 3.2

Assessing engagement as contributions to ARTful government

	Accountability, based on independence of true participants	Responsibility, based on representativeness of participants	Transparency of the process to the public
Referenda	High	Low-high, rising with turnout	High
Public hearings	Low	Low	Moderate
Public opinion surveys	High	High	Moderate
Negotiated rule making	Moderate	Low	Low
Consensus conferences	High	Moderate (limited by sample size)	High
Citizens' jury/panel	High	Moderate (limited by sample size)	Moderate
Citizens' advisory committee	Moderate	Moderate to low	Variable, often low
Focus group	High	Moderate (limited by sample size)	Low
Expert advisory group	Moderate	Moderate to high	Moderate to high

Source: Adapted from Rowe and Frewer (2000).

decision making. Third, processes must offer opportunities for face-to-face discussion over some period of time. Fourth, citizens should have some degree of equality with administrative officials and technical experts. Fiorino actually assesses five of the nine mechanisms. Extending his analysis using these criteria (see Table 3.3), one would argue that the role of amateurs and citizens tends to be accentuated but that sometimes this comes at the expense of a real share in the collective decision making or the loss of face time to debate the nuances of a system.

Fourth, the intensity of the engagement processes can be mapped against the public decision-making criteria to see at least where there is a fit between the nature of the engagement and the decision processes being used. Taking the Rowe and Frewer taxonomy, the International Association for Public Participation (IAP2) has defined a Public Participation Spectrum that describes five different levels of public involvement (IAP2 2007). The

TABLE 3.3

Assessing engagement through the "participation theory" lens

	Participation of amateurs	Share in collective decision making	Face-to-face discussion over time	Equality of citizens and experts
Referenda	Yes	Yes, if binding	No	Some
Public hearings	Yes	Limited	Varies	No
Public opinion surveys	Yes	Limited	No	No
Negotiated rule making	Unlikely	Yes	Yes	Yes
Consensus conferences	Yes	Limited	Yes	Varies
Citizens' jury/panel	Yes	Limited	Yes	Varies
Citizens' advisory committee	Yes	Limited	Yes	Some
Focus group	Yes	Limited	Varies	Yes
Expert advisory group	No	Varies based on mandate ·	Yes	No

Source: Adapted and expanded from Fiorino (1990).

spectrum (see Table 3.4) identifies the flow of information (from or to government), the degree of involvement, and the level of empowerment, differentiating between five levels of engagement, ranging from the lowest level (1), which involves only one-way flows of information from the government to citizens, to the highest level (5), with decision making in the hands of citizens.

For the intermediate levels (2, 3, and 4), where there are two-way flows of information between the government and the public, it is not clear how that information is to be used in decision making. Input from a wide range of stakeholders usually leads to conflicting opinions, and incorporating those differences in a decision can be challenging. For example, should the government consider the views of the majority, or should all actors compromise to reach a consensus? Pal and Maxwell (2004) offer a five-point framework that categorizes how governments might go about resolving conflicting input from various participants (see Table 3.5). First, decisions can be made or justified based on the process used; open, transparent, and inclusive systems can be used to justify a decision. Second, decisions can be based on the majority opinion, determined by referenda, votes, or polls. Third, a government might appeal to a utilitarian rationale, where the public interest is a

TABLE 3.4

Public participation spectrum

Level	Type	Characteristics	Examples
1	Inform	Provide balanced information to help understand issues and options	Fact sheets, websites, open houses
2	Consult	Obtain public feedback; acknowledge concerns and provide feedback on how public input influenced decisions	Public comments, focus groups, surveys, public meetings
3	Involve	Reflect public concerns directly in alternatives developed	Workshops, deliberative polling
4	Collaborate	Incorporate public advice and recommendations into decisions to the maximum extent possible	Consensus building
5	Empower	Put decision making in the hands of the public	Citizens' juries, ballots, delegated decisions

Source: IAP2 (2007).

balance or compromise of different interests based on the intensity of reactions (in an effort to avoid tyranny of the majority or sometimes tyranny of the loudest). Fourth, a government might seek a consensus on common pragmatic public interests. Fifth, the state can use shared values or normative principles to decide.

Combining these two concepts into a grid (see Table 3.6) offers us a way to see where there are logical fits. The scale and scope of engagement is represented in the rows, rising in each succeeding row. The decision-making systems are characterized in the columns, with each successive column to the right representing a more binding and engaged structure of decision making. As the table shows, the downward-sloping diagonal is most heavily populated with methods. As one might expect, if one is going to undertake decision making on a more normative basis (e.g., common interest and shared values), governments are going to need a more nuanced and deeper understanding of the values, beliefs, and interests of their citizens. As a reference point, both the vote and the individual purchase decision are included in the table since, in a perfect world, democracy and the market would be able to deliver them optimally. Clustered in the bottom

TABLE 3.5

Decision-making approaches

Approach	Description
Process	The public interest arises from and is served by fair, inclusive, and transparent decision-making procedures, and proper decisions will arise if proper procedures are followed. Proper procedures include legality, constitutionality, due process, transparency, fairness, equality of representation, and so on.
Majority opinion	The public interest is defined by what a significant majority of the population thinks about an issue. The majority opinion on a given issue can be determined by referenda, votes, or polls.
Utilitarian	The public interest is a balance or compromise of different interests involved in an issue, with the aim of maximizing benefits for society as a whole. This approach aims to overcome the tyranny of the majority phenomenon by accounting for the intensity of the reaction.
Common interest	The public interest is a set of pragmatic interests held in common, such as social stability, clean air, good water, defence, economy, and so on. This provides a mechanism for overcoming strongly vocal minority voices and even majority views if it is seen to be in the "true" public interest.
Shared values	The public interest is a set of shared values or normative principles. Values can be considered a good reflection of society at large since they are ingrained into the culture, and a political community cannot exist without some shared values. But such values might not be a good basis for specific policy actions since they might not recognize the diversity in contemporary Canada.

Source: Pal and Maxwell (2004).

centre as fully empowered models, neither is truly public engagement since they are done privately based on solely private motives. There is some evidence that processes are often used in circumstances that do not match up with the decision-making processes.

Fifth, Castle and Culver (2006) offer a reduced form typology, in which engagement involves "pushing" information out to citizens (analogous to Rowe and Frewer's communication), whereas public consultation requires both "pushing" information out and "pulling" information back into government. They posit that, to be morally valid, a public consultation must both

TABLE 3.6

Evaluation of democratic engagement by fit with decision rules

	Process	Majority	Utilitarian	Common interest	Shared values
Informing; giving information	Leaflets, newsletters, annual reports, Internet communications			Industry and stakeholder communications	Legislative hearings
Consulting; listening	Notification, distribution, and solicitation; public hearings	Citizens' panels; surveys; opinion polls	Qualitative interviews; focus groups; Delphi surveys; revealed preference studies	Delphi surveys; citizens' advisory panels	
Involving; visioning, exploring, innovating	Consultative workshops	Deliberative polling; non-binding referenda		Citizens' jury; consensus conference; dialogue tool	Sectoral or issue roundtables (e.g., Roundtable on the Environment and Economy)
Collaborating; judging or deciding together		Deliberative polls; referenda		Visioning	Joint venture between advocacy group and government (e.g., cancer foundations and CIHR)
Empowering; delegated decision making		Voting on binding referenda or propositions	Individual purchase decision		Communal trusts

Source: Based on criteria from IAP2 (2007) and Pal and Maxwell (2004).

intend to incorporate citizens' preferences and offer a binding obligation to use their authority to make good on the promise to use the input. Castle and Culver assessed two recent Canadian processes – the Canadian Biotechnology Advisory Committee (CBAC) consultation on the regulation of GM foods and a Genome Canada-funded research program examining opinions on GM salmon – and concluded that only the CBAC process met their standards for effective public consultation. Taking their two criteria and assessing the nine common processes, all of which have push and pull elements, and hence would be public consultation according to Castle and Culver, one can see (Table 3.7) that, though some forms of involvement might meet their first criterion (especially binding referenda and some of the intensive processes, such as conferences, panels, and groups, when they are conducted early in the policy process), only binding referenda involve binding obligations. Policy makers always reserve the right to accept or reject the output of any of these processes; in this way, they become instrumental

TABLE 3.7

Assessing the moral acceptability of public engagement

	Intent to incorporate citizens' preferences	Binding obligation to use authority
Referenda	Only if binding	Only if binding
Public hearings	Low	No
Public opinion surveys	Low-medium	No
Negotiated rule making	Low-high, depending on citizen engagement	Usually high
Consensus conferences	Varies; higher at earlier stages of policy analysis	Unlikely
Citizens' jury/panel	Varies; higher at earlier stages of policy analysis	Unlikely
Citizens' advisory committee	Varies; higher at earlier stages of policy analysis	Variable, often low
Focus group	Varies; higher at earlier stages of policy analysis	No
Expert advisory group	Low-high; higher at earlier stages of policy analysis	Varies; depends on mandate

Source: Criteria from Castle and Culver (2006).

touchstones for policy decisions but seldom fundamentally drive policy outcomes.

The five somewhat distinct evaluations of nine common processes of democratic engagement offer some insights. First, none completely maps onto the democratic norms; hence, one cannot offer unambiguous advice to policy makers on which ones to pick to solve that dilemma. Second, moving beyond the nature of the system to its outputs, if we want reflexive decisions that are accountable, responsible, and transparent, we will be equally challenged. Rowe and Frewer's framework shows that none can deliver all three outcomes – policy makers need to choose among them or use more than one mechanism and then figure out how to reconcile their differences (as suggested by Fiorino 1990). Third, Fiorino's criteria suggest that, though we might be engaging amateurs, we have yet to design a single model that satisfies the other important criteria. Fourth, a relational mapping between the degree of engagement and decision-making criteria reveals that there are natural pairings of types of engagement and specific decision systems. Fifth, Castle and Culver's two criteria reveal that, at least as used today, none of the public consultation methods is ethically grounded, mostly because those controlling the levers of power refuse to meet their second criterion of binding themselves to using public perceptions in decision making.

Conclusion: Complexity, Reflexivity, and ARTful Government

Political scientist James Rosenau (1995, 16) asserts that global governance is "the sum of myriad – literally millions of – control mechanisms driven by different histories, goals, structures, and processes." Paquet (2001, 190) suggests that we face new forms of distributed governing arrangements "based on a more diffused pattern of power" in which distributed governing "does not simply mean a process of dispersion of power toward localised decision-making within each sector: it entails a dispersion of power over a wide variety of actors and groups." A seemingly infinite array of distributions of power evolves in response to pressures to adjust to rapid change. "The fabric of these new 'worlds' is defined by the new dominant logic of subsidiarity in all dimensions: it welds together assets, skills and capabilities into complex temporary communities that are as much territories of the mind as anything that can be represented by a grid map" (190).

Economist Herbert Simon has examined the concept of complexity, concluding that the interest in recent years of many sciences in complexity and complex systems has drawn attention to the fact that most of the complex systems seen in the world are nearly decomposable systems. They are arranged in levels, the elements at each lower level being subdivisions of the elements at the level above. Molecules are composed of atoms, atoms of electrons and nuclei, electrons and nuclei of elementary particles. Multi-celled organisms are composed of organs, organs of tissues, tissues of cells ... Near decomposability is a means of securing the benefits of coordination while holding down its costs by an appropriate division of labour among subunits ... Nearly decomposable systems will adapt to the changing environment and gain in fitness more rapidly than systems without this property. (2000, 793)

Although one might argue that our system should become more reflexive as it fragments, that is not a foregone conclusion. Reflexivity might ultimately be incompatible with the truly accountable and responsible state institutions (be they international, national, or subnational), and the more that is left to the efficient market, the less room there will be for introspection. Although civil authorities appear to be positioned to articulate the views of groups in society – both for-profit groups that self-regulate and manage market functions such as pre-commercial, non-competitive research, standards setting, and market development and not-for-profit social action groups that represent segments of society that are not adequately addressed by private or public actors – few are truly ARTful. Many are often only vehicles for special interests. Large swaths of public opinion are not mobilized. Political scientist Robert Gilpin (2001) has suggested that, at the extreme, there is a risk that a nearly decomposable system might lead to a neomedieval system in which we are overgoverned by a complex web of vested interests, analogous to the guild structure of the Middle Ages.

The challenge is to avoid that outcome. Democratic engagement could be part of the solution – or simply accentuate the problem. As noted above, there has been a tendency to jump quickly to the conclusion that democratic engagement is the answer, without clearly setting up the rationale for the problem that it is intended to fix. Although this chapter has attempted to provide a framework to link the problem to the solution and to test whether the solution actually addresses and ameliorates the problem, more detailed work is needed to find a solution that better fits the problem.

NOTES

This research has been supported by VALGEN (Value Generation through Genomics and GE³LS), a project supported by the Government of Canada through Genome Canada and Genome Prairie.

1 See Polity IV Project, http://www.systemicpeace.org/.

REFERENCES

Arrow, K. 1951. *Social Choice and Individual Values.* New York: Wiley.

Bentley, A.F. 1908. *The Process of Government: A Study of Social Pressures.* Chicago: University of Chicago Press.

Boulding, K. 1970. "Organizers of Social Evolution." In *A Primer on Social Dynamics: History as Dialectics and Development,* edited by K. Boulding, 19-36. London: Collier Macmillan.

Brown, Eric. 2009. "Plato's Ethics and Politics in *The Republic.*" *Stanford Encyclopedia of Philosophy* (winter 2011 edition). http://plato.stanford.edu/archives/.

Castle, D., and K. Culver. 2006. "Public Engagement, Public Consultation, Innovation, and the Market." *Integrated Assessment Journal: Bridging Sciences and Policy* 6, 2: 137-52.

Dahl, R. 1998. *On Democracy.* New Haven: Yale University Press.

Downs, A. 1957. *An Economic Theory of Democracy.* New York: Harper and Row.

Fiorino, D. 1990. "Citizen Participation and Environmental Risk: A Survey of Institutional Mechanisms." *Science, Technology, and Human Values* 15: 226-43.

Gilpin, R. 2001. *Global Political Economy: Understanding the International Economic Order.* Princeton: Princeton University Press.

Hardin, R. 2002. "Street-Level Epistemology and Democratic Participation." *Journal of Political Philosophy* 10, 2: 221-29

Held, D., A. McGrew, D. Goldblatt, and J. Perraton. 1999. *Global Transformations: Politics, Economics, and Culture.* London: Polity.

Howlett, M., M. Ramesh, and A. Pearl. 2009. *Studying Public Policy: Policy Cycles and Policy Subsystems.* 3rd ed. Toronto: Oxford University Press.

International Association of Public Participation (IAP2). 2007. IAP2 Spectrum of Public Participation. http://www.iap2.org/.

Olson, M. 1965. *The Logic of Collective Action: Public Goods and the Theory of Groups.* Cambridge, MA: Harvard University Press.

Osborne, D., and T. Gaebler. 1992. *Reinventing Government: How the Entrepreneurial Spirit Is Transforming the Public Sector.* London: Plume-Penguin.

Pal, L., and J. Maxwell. 2004. "Assessing the Public Interest in the 21st Century: A Framework." Paper prepared for the External Advisory Group on Smart Regulation (Canada). http://rcrpp.org/.

Paquet, G. 2001. "The New Governance, Subsidiarity, and the Strategic State." In *Governance in the 21st Century,* 183-208. Paris: OECD. http://www.oecd.org/.

Phillips, P. 2007. *Governing Transformative Technological Innovation: Who's in Charge?* Oxford: Edward Elgar.

Picciotto, R. 1995. "Putting Institutional Economics to Work: From Participation to Governance." World Bank Discussion Paper 304. http://www-wds.worldbank. org/.

Putnam, R. 2000. *Bowling Alone: The Collapse and Revival of American Community.* New York: Simon and Schuster.

Rhodes, R. 1995. *The New Governance: Governing without Government.* State of Britain Seminar II of a joint ESRC/RSA seminar series, 24 January.

Rosenau, J. 1995. "Governance in the Twenty-First Century." *Global Governance* 1: 13-43.

Rowe, G., and L. Frewer. 2000. "Public Participation Methods: A Framework for Evaluation." *Science, Technology, and Human Values* 25, 1: 3-29.

–. 2005. "A Typology of Public Engagement Mechanisms." *Science, Technology, and Human Values* 30, 2: 251-90.

Sen, A. 1979. "Personal Utilities and Public Judgements: Or What's Wrong with Welfare Economics." *Economic Journal* 89: 537-88.

Simon, H. 2000. "Public Administration in Today's World of Organizations and Markets." *Political Science and Politics* 33, 4: 749-56.

Stoker, G. 1998. "Governance as Theory: Five Propositions." *International Social Science Journal* 50, 155: 17-28.

Truman, D. 1951. *The Governmental Process: Political Interests and Public Opinion.* London: Greenwood Press.

PART 2

EXTERNAL CONDITIONS FOR LEGITIMATE PUBLIC ENGAGEMENT
Ethics, Society, and Democracy

4 Trust, Accountability, and Participation
Conditions for and Constraints on "New" Democratic Models

SUSAN DODDS

This chapter explores the relationships of trust (specifically public trust in scientific institutions and government regulators), social epistemology concerning science and its significance, and the democratic ideals of deliberation through public exchange. I argue that, because of the epistemic context in which public deliberation about policy concerning developing technologies occurs, policy makers require both lay public and expert contributions to knowledge. The legitimacy of dependence on expert authority in identifying relevant considerations in the debate and in responding to concerns or hopes raised by lay publics will depend, in part, on the well-founded trust of those publics in the relevant experts. That trust cannot be secured simply through the constraint of accountability mechanisms on experts. As a result, though public engagement processes *might* contribute to public trust in science and policy makers, public involvement in discussions of developing technologies cannot *create* the conditions for the legitimacy of public policy outcomes that occur when trust is either absent or unwarranted. Legitimate public deliberation on emerging technologies and the development of public policy through public deliberative processes depend on a number of conditions, of which a threshold of well-founded trust in policy makers and technological experts is one part.

Public engagement in discussions on appropriate policy concerning novel technologies, it is argued, promotes the democratic values of political reasoning, accountability, and non-domination. Public participation does so by drawing publics into understanding of, and reasoning about, matters that can affect wide populations; by challenging the authority of technocrats (both scientific and regulatory) to provide responses that address the array of articulated public concerns; and by reducing the ability of those with greater access to epistemic or material resources to influence public policy on matters that are (or are thought should be) of wider public concern.

Nonetheless, many of the people who endorse public engagement as a means of achieving democratic ends are also wary of the misuse of these events and of the possibility that policy makers' enthusiasm for public engagement will be abused or misplaced in the absence of a real commitment to the value of participants' knowledge (Irwin 2006; Irwin and Wynne 1996; Wynne 2006). Brian Wynne, for example, notes the use of a range of public engagement strategies as "moves to 'restore' public trust in science by developing an avowedly two-way, public dialogue with science initiatives," but then he shows that, in many cases, the dialogic aspirations of public engagement revert to the view that the public lacks understanding of science and that expert positions, properly understood, would be endorsed by publics through the engagement process. He argues that there is as much work to be done to redress the "scientific deficits of understanding of publics" as has been done to redress the claimed "deficits" in the public understanding of science (2006, 216). Similarly, where publics engage in deliberation but subsequently find that their participation has had no effect on policy or practice (i.e., where the deliberation does not influence decision makers), the effect on public confidence in the technology and policy makers can be worse than where the public has no voice at all: "In the absence of real influence, the illusion of voice can lead to even greater frustration and disenchantment than having no voice at all" (Delli Carpini, Cook, and Jacobs 2004, 333). Public trust can be fostered, abused, or lost through public engagement concerning developing technologies, implying that it should be understood that trust and its conditions are vital to any process of engagement.

Deliberation, Pragmatism, and Democracy
The current political interest in the use of a range of "public talk" processes to secure public support for developments in science policy (among other areas) parallels developments in democratic theory that emphasize the engagement of publics in political processes of public reasoning, including works on participatory democracy and republican theory (Dagger 1997; Young 2000), public justification (Gaus 1996), and deliberative democracy

(Dryzek 2000; Elster 1998; Goodin 2003; Gutmann and Thompson 1996; Habermas 1996). These theorists share a concern to shift from "mere aggregation" of preferences reflected in voting procedures to a more complex mix of talk, dialogic reason giving and reflection, expression of the diversity of viewpoints, mutual respect, and deliberation on a shared outcome (not necessarily supported by consensus). Theorists working in this area view deliberative approaches as offering responses to the apparent depoliticization of the citizenry in advanced capitalist democracies in which the expansion of the market has displaced civic engagement. Public participation and direct public accountability are aimed at redressing diminishing trust in public institutions and the demise of the commitment to civic virtue by both citizens and public officials. Deliberative democrats seek to restore legitimacy to democracy by restoring a collectivist notion of the "general will" authorizing political authority.

For Habermas (1996, 28), the justification of political decisions is found in the "procedures and communicative presuppositions of democratic opinion- and will-formation." Deliberative approaches to democracy emphasize the legitimation of policy that comes from the transformation of interests through processes of "collective decision making by all those who will be affected by the decision or their representatives: this is the democratic part. Also ... it includes decision making by means of arguments offered by and to participants who are committed to the values of rationality and impartiality: this is the deliberative part" (Elster 1998, 8).

Within deliberative approaches to democracy, public engagement or deliberative events play a role in the democratization of knowledge by demanding of experts that they give an account of their reasoning and recommendations on policy that, in principle, citizens could endorse (Chambers 2003; Delli Carpini, Cook, and Jacobs 2004). That is, citizens could consent to being subject to the policy because, though they might lack the expertise to judge the technical matters, the reasoning supporting the policy is comprehensible, citizens have been given an opportunity to challenge or question the experts on the matter, and the expert reasoning is responsive to the concerns or values that the citizens have expressed as shaping the context within which the policy can be developed if it is to have their endorsement.

Noting that actual deliberative processes occur outside Habermas's "ideal speech" situations, Fishkin (1995, 41) argues that deliberative processes exist on a continuum of completeness (of reason giving, of participants' preparedness to consider arguments presented, of available information, etc.)

that is related to the legitimacy (understood as accountability) of a deliberative process:

> When arguments offered by some participants go unanswered by others,
> when information that would be required to understand the force of a claim
> is absent, or when some citizens are unwilling to weigh some of the argu-
> ments in the debate, then the process is less deliberative because it is
> incomplete in the manner specified. In practical contexts a great deal of
> incompleteness must be tolerated. (cited in Delli Carpini, Cook, and Jacobs
> 2004, 317)

Where deliberators are faced with very incomplete processes, they cannot
in principle endorse the reasoning supporting the policy and hence cannot
give it their consent.

One source of the "deliberative turn" in political theory is an earlier
democratic tradition, the pragmatism of the late-nineteenth-century and
early-twentieth-century philosophers James, Peirce, and Dewey and, particularly, Dewey's democratic theory. The pragmatists can be said to argue
for the maxim that "we clarify a hypothesis by identifying its practical
consequences" (Hookway 2008, n.p.). Where policy matters are the subject of inquiry, the identification of practical consequences of hypothetical
responses is completed through discussion, consultation, persuasion, and
debate. Dewey importantly viewed democracy as a form of social inquiry:
problem solving through public deliberation (Festenstein 2005). Gastil
(2000, 22) sees Dewey's understanding of full deliberation as involving "a
careful examination of a problem or issue, the identification of possible
solutions, the establishment or reaffirmation of evaluative criteria, and
the use of these criteria identifying an optimal solution." Bohman (1999,
590) notes that, for the pragmatists, "deliberative democracy must have at
least two aspects: not only free and open debate and discussion, but also
socially organized deliberation on how best to achieve effective consensual ends."

Science and democracy (as an experimental social science) require a
"community of inquiry" governed by consensus on norms of evidence.
However, given that some expert knowledge might be necessary to achieve
the best evidence and an optimal solution to a given problem, and that "no
single person, expert or lay, fully understands all of the intricacies of any
specific decision" (Bohman 1999, 591), there is the need for an epistemic
division of labour, combined with robust processes for public deliberation.

This democratic inquiry brings the reasoning of experts into deliberative evaluation alongside lay public values, concerns, and understanding of what is at stake in relation to the policy in question. "These processes extend and deepen the public awareness of the problems under discussion, and help to inform the 'administrative specialist' of social needs" (Festenstein 2005, n.p.).

For the pragmatist, the pursuit of knowledge is inherently social and dialogic: it is through the process of reasoning and testing of reasons in relation to their practical consequences that we are able to make claims of knowledge or claims that a particular hypothesis has been established. Furthermore, because no one is an expert on all matters, knowledge is socially distributed. "If inquiry is democratically organised, then socially distributed knowledge is not represented anywhere but in the group as a whole" (Bohman 1999, 594).

Trust and Epistemology

Pragmatists hold the view that any individual's claims to knowledge depend on the social distribution of expertise and the public testing of knowledge claims. As a result, each individual must rely on the effectiveness of the social norms of reasoning and the accountability of experts in forging their own interests based on that knowledge. This raises the problem of *trust* in large and complex societies (Bohman 1999, 592). According to Bohman, there is a tension between the epistemic efficiency of relying on the authority of experts and the democratic value of public inquiry by citizens to determine the terms of social cooperation. This tension can be posed as one between the quality of political decisions and their legitimate authority (see Estlund 2008). Bohman argues that the tension is resolved through democratic deliberative processes that address the *credibility* of expert authority and the *legitimacy* of norms of cooperation between experts and lay publics. Trust in expert claims to knowledge is democratic where two conditions are filled: "It must establish free and open interchange between experts and the lay public and discover ways of resolving recurrent cooperative conflicts about the nature and distribution of social knowledge" (592).

Justified trust in experts, then, is not blind acceptance of the epistemic authority of the scientist. Rather, it is a trust founded on a critical evaluation of the norms and institutions that support the claims of expertise; deference to the authority of the expert is democratically justified where the claim to expertise on the issue is acceptable to citizens engaged in the project of social cooperation.

Skepticism about whether trust in expert authority is warranted is found in Irwin's (1995) calls for critical public scrutiny of scientific expertise and for experts to engage in open debate on the value and risks of relevant scientific developments so that citizen deliberators are in a position to weigh the merits of the scientific arguments for themselves. Similarly, Dryzek (2000, 165) argues that an effective citizen voice in economic and technological developments requires a citizenry that is appropriately circumspect in its response to the authority of experts and that the citizenry should withhold trust where it has reason to doubt that it is well placed:

> Distrust of experts does not mean that everyone has to become an expert. Instead, it can mean approaching expert testimony with a sceptical attitude, perhaps questioning the credentials of experts, seeking corroboration for any contentious claim, refusing to believe an expert if his or her research is funded by the offending industry, or if his or her record indicates an axe to grind.

The authority of experts in democratic deliberation depends on their ability to foster and secure well-founded public trust through their demonstrated accountability for the positions that they defend in democratic deliberative fora. However, the skepticism about the trustworthiness of experts reflected by Irwin's and Dryzek's arguments suggests that deliberative public reasoning might ultimately fail in the absence of pre-existing confidence in information providers and policy makers.

Trust and Hope as Conditions for Effective Deliberation

In the previous section, I argued that the legitimacy of policy decisions reached through deliberation depends, in part, on the accountability of deliberators and policy makers who transform public reasoning on a policy matter into a resolution. However, the prospect that deliberators, and especially experts, will be held accountable for their advice is not sufficient to protect citizens against the political domination of experts, nor is it sufficient to ensure the legitimacy of policy based on deliberation governed by norms of accountability. The absence of trust between deliberators and experts or policy makers can threaten the effectiveness of deliberation in eliciting knowledge or understanding, it can threaten the accountability of experts or policy makers in responding to the concerns and arguments of participants, and it can therefore threaten the legitimacy of any policy outcome generated by the deliberative process.

Citizens who believe that they are vulnerable to domination or subordination to expert claims to knowledge can withhold their willingness to participate as democratic deliberators, because they do not experience the social distribution of knowledge as serving democratic ends. As a result, such citizens will not endorse the outcome of the deliberative process as legitimate. Furthermore, citizens who distrust policy makers to make policies that are respectful of the values and concerns of the citizenry can withhold their contributions to the deliberative process, refusing to offer up their values, interests, and concerns to the test of public reasoning if they lack trust in the process. In this case, the policy deliberations will lack relevant information against which to evaluate and test policy alternatives, and they will fail to respond to the actual interests and concerns of the citizenry. Because participation in any deliberative process requires participants (especially those already most vulnerable to the effects of a policy outcome) to make themselves vulnerable to the democratic process, it requires as a precondition the trust of deliberators in the deliberative process to effectively engage their reason and be politically legitimate.

A further level of complexity arises from the reflexive nature of trust and the tri- (or multi-) partite nature of public policy deliberation about emerging technologies. For a deliberative process to successfully engage the reasoning of all parties, the diverse lay publics need to hold a threshold of trust in the experts who interpret the significance of lay public issues and concerns in relation to expert knowledge, and they must have sufficient trust in policy makers to use the interpreted information to which they have contributed in a manner that will not harm the interests of the lay public. Furthermore, to understand themselves as able to participate effectively as citizens in contributing to the deliberation, lay participants need to have a level of self-trust and self-respect (Anderson and Honneth 2005). Similarly, for experts to faithfully interpret and engage with lay publics' reasoning and concerns, they need to trust and respect participants; to faithfully reflect that input into policy deliberations, they also need to place trust in policy makers to use their findings appropriately. Experts also need to have confidence in their own judgments (self-trust) and self-respect. The third party, policy makers, similarly needs to trust lay public participants, experts, and their own authority and political expertise for the deliberation to be effective.

But what is trust, and how is warranted trust to be understood? At the interpersonal (rather than institutional) level, "trust is an attitude that we have towards people whom we hope will be trustworthy, where trustworthiness is a property, not an attitude" (Macleod 2006, n.p.). Trust requires that the person who trusts (the truster) accepts her vulnerability to the person trusted (the trustee), because the trustee might not "pull through" for the truster. Or, as Mark Warren (1999, 311) puts it, "trust ... involves a judgement, however tacit or habitual, to accept vulnerability to the potential ill will of others by granting them discretionary power over some good." The attitude of trust is more likely to occur where the truster accepts the risk that trusting entails, including the risk of betrayal, where the truster is inclined to expect the best of the trustee, and where the truster has a level of confidence in the competence of the trustee (at least with regard to the entrusted realm). According to Baier (1994), trust also involves the belief that the trustee is motivated by goodwill toward the truster; according to Jones (1996), the trustee is moved by the thought that the truster is counting on her. Trust is warranted (or justified) when the truster is justified in believing (or hoping) that the trustee is trustworthy. However, because trust invariably involves vulnerability, it must be extended beyond cases where the truster has sufficient evidence of trustworthiness to guarantee that trust is warranted (I don't have to trust you when I am certain that you cannot betray my trust; I must trust you when I can't force you to make good on my trust). Hence, McGeer (2008, 240) describes *substantive trust* (or what Govier [1997] refers to as *thick trust*) as trust that has two related features: "(1) it involves making or maintaining judgements about others, or about what our behaviour should be towards them, that go beyond what the evidence supports; and (2) it renounces the very process of weighing whatever evidence there is in a cool, disengaged, and purportedly objective way." For McGeer, the challenge is to explain why substantive trust can be rational. She argues that substantive trust is rational insofar as our hopes for the trustee provide a kind of active affective "scaffolding" that brings out from the trustee the hoped-for actions: "In trusting others and so hoping for their trust-responsive care and competence, we ask something substantial of them ... By way of such hopeful scaffolding, we also give trusted others something substantial in return – namely, a motivationally energizing vision of what they can do or be" (249). For McGeer,

> Our hopeful investment of trust in others can often elicit – or, better, empower – trust-responsive behaviour of the sort we seek: namely, acts and attitudes on the part of trustees that live up to our hopeful vision of what they can do and be, particularly with regard to showing competence and care in the domain in which we trust them. (250)

Thus, in the interpersonal case, trust is supported by the capacity for our hopeful investment in others to give trustees the agential capacity to "pull through" for us. Although such optimism in the capacities and motivations of others can be misplaced, or disappointed, it can also draw out of trustees the best that they can be and justify the trust of the truster.

Trust and distrust do not cover the range of attitudes that people can have toward one another or toward institutions and experts. The absence of trust does not imply distrust – disinterest can frequently characterize our attitudes to those whom we neither trust nor distrust, where those people are not in a position to harm our interests, and where we are unlikely to be reliant on them. Distrust shares some characteristics with trust: it is the attitude that we have toward those we fear are untrustworthy, it is resilient (once our trust is lost, it is hard to regain), it often extends beyond evidence of untrustworthiness, it is shaped by knowledge and experience, and it is an attitude that shapes the epistemic uptake of the claims of the person who is distrusted (Govier 1997; McGeer 2008; Warren 1999). Where an individual has come to distrust another, especially where the distrust is substantive, it is particularly resilient. For this reason, one cannot assume that deliberative democratic processes, even where accountable and respectful of participants' knowledge, will foster sufficient trust among participants who have become distrustful.

So how far can we take the analysis of the interpersonal case in assessing the rationality of trust in collectives or institutions? As Hardin (1999, 23) notes, modern political communities are too complex for citizens' trust in governments to be warranted most of the time: "In general, citizens cannot know enough of what they must know in order to be able to trust government." Our disposition to trust systems or institutions (rather than individuals) depends on a number of additional factors, including the social circumstances of the truster (ourselves) and trustee (institution or system), the number and diversity of sources of knowledge about the matter being entrusted and the trustee's reliability, the systems of accountability that frame the trustee institution's authority, and the degree to which we believe that the institutional systems of accountability are framed so as to protect or promote our interests (see Govier 1997). Not surprisingly, members of groups that have had negative or prejudicial experiences of social systems or institutions feel the most vulnerable to those institutions and are least likely to invest their trust in those systems, even when the individuals who represent the institutions are trustworthy, motivated by concern and care for others, accountable, competent, and reliable. Fostering trust in public institutions takes more than a few trustworthy souls; it takes evidence of the trustworthiness of the system and its processes of accountability.

Conclusion

Creating the conditions for democratic legitimacy through public participation requires more than the establishment of deliberative decision-making processes that engage publics. Those who have touted the value of public participation as a means of generating public trust in science, for example, might well have grasped the wrong end of the stick; public participation in deliberative democratic decision making as a means of legitimating policy outcomes depends on there already being (warranted) substantive trust in the democratic institutions that invite public deliberation. As Warren (1999, 340) notes, it is easier to recognize trust and deliberative decision making as complementary to democratic theory than it is to claim that political deliberation produces public trust. Although opportunities for robust public reasoning can contribute (if properly handled to respectfully honour the trust of participants) to the level of institutional trust over time, they cannot generate trust where there is none and might serve to erode already faltering public trust.

NOTE

This research was supported by the Australian Research Council Discovery Grant DP0556068, Big Picture Bioethics: Policy-Making and Liberal Democracy, and the University of Tasmania.

REFERENCES

Anderson, J., and A. Honneth. 2005. "Autonomy, Vulnerability, Recognition, and Justice." In *Autonomy and the Challenges to Liberalism*, edited by J. Christman and A. Anderson, 127-49. Cambridge, UK: Cambridge University Press.

Baier, A. 1994. "Trust and Its Vulnerabilities." In A. Baier, *Moral Prejudices: Essays on Ethics*, 130-51. Cambridge, UK: Cambridge University Press.

Bohman, J. 1999. "Democracy as Inquiry, Inquiry as Democratic: Pragmatism, Social Science, and the Cognitive Division of Labor." *American Journal of Political Science* 44, 2: 590-607.

Chambers, S. 2003. "Deliberative Democratic Theory." *Annual Review of Political Science* 6: 307-26.

Dagger, R. 1997. *Civic Virtues: Rights, Citizenship, and Republican Liberalism*. New York: Oxford University Press.

Delli Carpini, M.X., F.L. Cook, and L.R. Jacobs. 2004. "Public Deliberation, Discursive Participation, and Citizen Engagement: A Review of the Empirical Literature." *American Review of Political Science* 7: 315-44.

Dryzek, J.S. 2000. *Deliberative Democracy and Beyond: Liberals, Critics, Contestations.* New York: Oxford University Press.

Elster, J. 1998. "Introduction." In *Deliberative Democracy,* edited by J. Elster, 1-18. Cambridge, UK: Cambridge University Press.

Estlund, D.M. 2008. *Democratic Authority: A Philosophical Framework.* Princeton: Princeton University Press.

Festenstein, M. 2005. "Dewey's Political Philosophy." *Stanford Encyclopedia of Philosophy* (spring 2009 edition). http://plato.stanford.edu/archives/.

Fishkin, J.S. 1995. *The Voice of the People: Public Opinion and Democracy.* New Haven: Yale University Press.

Gastil, J. 2000. *By Popular Demand: Revitalizing Representative Democracy through Deliberative Elections.* Berkeley: University of California Press.

Gaus, G.F. 1996. *Justificatory Liberalism: An Essay on Epistemology and Political Theory.* New York: Oxford University Press.

Goodin, R.E. 2003. *Reflective Democracy.* New York: Oxford University Press.

Govier, T. 1997. *Social Trust and Human Communities.* Montreal: McGill-Queen's University Press.

Gutmann, A., and D. Thompson. 1996. *Democracy and Disagreement.* Cambridge, MA: Harvard University Press.

Habermas, J. 1996. "Three Normative Models of Democracy." In *Democracy and Difference: Contesting the Boundaries of the Political,* edited by S. Benhabib, 21-30. Princeton: Princeton University Press.

Hardin, R. 1999. "Do We Want Trust in Government?" In *Democracy and Trust,* edited by M. Warren, 22-41. Cambridge, UK: Cambridge University Press.

Hookway, C. 2008. "Pragmatism." *Stanford Encyclopedia of Philosophy* (spring 2010 edition). http://plato.stanford.edu/archives/.

Irwin, A. 1995. *Citizen Science: A Study of People, Expertise, and Sustainable Development.* New York: Routledge.

–. 2006. "The Politics of Talk: Coming to Terms with the 'New' Scientific Governance." *Social Studies of Science* 36: 299-320.

Irwin, A., and B. Wynne, eds. 1996. *Misunderstanding Science? The Public Reconstruction of Science and Technology.* Cambridge, UK: Cambridge University Press.

Jones, K. 1996. "Trust as an Affective Attitude." *Ethics* 107: 4-25.

Macleod, C. 2006. "Trust." *Stanford Encyclopedia of Philosophy* (spring 2011 edition). http://plato.stanford.edu/archives/.

McGeer, V. 2008. "Trust, Hope, and Empowerment." *Australasian Journal of Philosophy* 86, 2: 237-54.

Warren, M. 1999. "Democratic Theory and Trust." In *Democracy and Trust,* edited by M. Warren, 310-45. Cambridge, UK: Cambridge University Press.

Wynne, B. 2006. "Public Engagement as a Means of Restoring Public Trust in Science: Hitting the Notes, but Missing the Music?" *Community Genetics* 9: 211-20.

Young, I.M. 2000. *Inclusion and Democracy.* New York: Oxford University Press.

Public Voices or Private Choices?
The Role of Public Consultation in the Regulation of Reproductive Technologies

COLIN GAVAGHAN

In countries such as the United Kingdom, New Zealand, and Canada, the regulation of assisted reproductive technologies (ARTs) takes place "against a backcloth of ethical pluralism" (Brownsword 2009, 40). In that context, it is a significant challenge for regulators to take account of the myriad ethical principles thereby engaged. This task is rendered no easier by the fact that widely shared principles such as "autonomy," "beneficence," and "justice" are open to numerous definitions, to say nothing of the relatively new kids on the bioethical block, such as "dignity" (Beyleveld and Brownsword 2001), "humility" (Sandel 2007), and "authenticity" (Brownsword 2008).

It is incumbent on those regulators, however, to set about this task seriously. In particular, those charged with implementing, interpreting, and enforcing what Julia Black (2001) calls "command and control" regulation – that backed with criminal legal sanctions – relating to the most important and intimate areas of people's lives owe those people a clear explanation of their decisions.

That explanation, I suggest, should derive from widely shared moral values, clearly understood and consistently applied. In determining which principles they may be, it is undeniable that public consultation can play a vital role. However, it is equally true that public opinion can be used as an excuse to preclude choices that are merely unusual or unpopular or as a mechanism by which a regulatory body can abdicate its responsibility of ethical scrutiny.

The manner in which pre-implantation genetic diagnosis (PGD), and, in particular, pre-implantation sex selection, have been regulated in the United Kingdom, I will argue, provides an example of the misuse of public consultation; the consultative process was used to substitute simple majoritarianism for reasoned deliberation; the consultation became a de facto referendum rather than any sort of exercise in truly *deliberative* democracy.

Equally importantly, the regulatory body's response to the consultation – a response later mirrored in legislation – failed to engage with important jurisprudential and ethical concerns. In particular, it failed to consider whether this technique, and the choices that it allows, should properly be considered to fall within the "realm of private decision making" that is often thought to be a prerequisite of any liberal democracy or whether they transgress against a shared social morality in a genuine and significant manner.

The Liberal Presumption and the Harm Principle

The notion that individual liberty ought to be protected, even in the face of mass opinion, can be traced back at least as far as the nineteenth-century political philosopher John Stuart Mill. In "On Liberty," probably his most impassioned work, Mill ([1859] 1979, 129) railed against "the tyranny of the majority," claiming that "there is a limit to the legitimate interference of collective opinion with individual independence." In the essay's best-known passage, Mill set out a theory with regard to the extent of this "legitimate interference":

> The sole end for which mankind are warranted, individually or collectively, in interfering with the liberty of action of any of their number, is self-protection. That the only purpose for which power can rightfully be exercised over any member of a civilised community, against his will, is to prevent harm to others. His own good, either physical or moral, is not a sufficient warrant.

Mill's argument placed the onus on those who would restrict the conduct of others to advance a justification in terms of harm. Absent such a justification, the state had no business imposing itself on the choices of its citizens.

A similar approach informed the recommendations of the United Kingdom's Committee on Homosexual Offences and Prostitution, immortalized in the study of contemporary ethics as the Wolfenden Committee (Committee on Homosexual Offences and Prostitution 1957). The Committee's famous report was published at a time when homosexual acts between

consenting adult males were criminal offences, often leading to imprison-
ment. In the event, the report recommended not only the decriminalization
of consensual, adult homosexual acts but more generally that "there must
remain a realm of private morality and immorality which is, in brief and
crude terms, not the law's business" (Wolfenden Committee 1957, 24). This
claim – and its possible implications – formed the starting point for what
Peter Cane (2006, 22) has described as "one of the most important jurisprudential debates of the second half of the 20th-century."

The notion that the legitimate province of criminal law lies in the preven-
tion of harm did not, of course, go unchallenged. Patrick Devlin (1965), a
member of the English judiciary, presented an influential contrary argument on the basis of a legitimate right for a society to defend its moral values
against transgressors, even when those transgressors pose no threat of actual harm to anyone. A society, Devlin claimed, is predominantly constituted not by tangible structures such as courtrooms and schools but by
shared values: "For society is not something that is kept together physically;
it is held by the invisible bonds of common thought. If the bonds were too
far relaxed the members would drift apart. A common morality is part of
the bondage. The bondage is part of the price of society; and mankind,
which needs society, must pay its price" (10). That being so, the upholding
of those values – even by force of law if required – can be seen as a form of
moral self-defence: "Society may use the law to preserve morality in the
same way as it uses it to safeguard anything else that is essential to its existence." Devlin recognized that it is neither possible nor desirable to prevent
every act offensive to anyone else's personal moral code. Rather, he maintained, "there must be toleration of the maximum individual freedom that
is consistent with the integrity of society ... Nothing should be punished by
the law that does not lie beyond the limits of tolerance" (16-17).

A number of influential legal theorists rallied around the liberal thesis of
the Wolfenden Committee's report. Most famously, Herbert Hart (1963), in
his series of lectures at Stanford University and collected in *Law, Liberty,
and Morality*, adopted a stance against what he referred to as "legal moral-
ism." Echoing Mill's "tyranny of the majority," Hart warned against "moral
populism," which he saw as "a misunderstanding of democracy which still
menaces individual liberty" (79).

Ronald Dworkin, who became another leading figure in liberal jurisprudence, also came to the Wolfenden Committee's defence, arguing for "our
tradition of individual liberty, and our knowledge that the morals of even
the largest mob cannot come warranted for truth" (1966, 986). Much of his

argument relied on an attempt to distinguish genuine moral perspectives – which a conscientious legislator should certainly take into account – from other states of mind – to which that legislator was required to pay no heed. Dworkin maintained that Devlin used the notion of "societal morality" in an "anthropological sense," concerning himself with the number of members in his society who thought a certain way rather than the quality of their opinions:

> Even if it is true that most men think homosexuality an abominable vice and cannot tolerate its presence, it remains possible that this common opinion is a compound of prejudice (resting on the assumption that homosexuals are morally inferior creatures because they are effeminate), rationalization (based on assumptions of fact so unsupported that they challenge the community's own standards of rationality), and personal aversion (representing no conviction but merely blind hate rising from unacknowledged self-suspicion) ... If so, the principles of democracy we follow do not call for the enforcement of the consensus, for the belief that prejudices, personal aversions and rationalizations do not justify restricting another's freedom itself occupies a critical and fundamental position in our popular morality. (1000-1)

As Cane (2006, 29) notes, however, "whatever one thinks ... of Devlin's views about the limits of the criminal law, he cannot plausibly be accused of holding the opinion that whether homosexual activity (or any other type of conduct) should be (de)criminalized depended merely on facts about what British people in the 1950s and 1960s actually believed."

Rather, for his "limits of tolerance" to be reached, Devlin (1965) seemed to require both a qualitative and a quantitative element to the intolerance. With regard to the former, "it is not nearly enough to say that a majority dislike a practice; there must be a real feeling of reprobation" (16-17). He referred several times to "disgust" and "indignation" as indicators that the limits of tolerance were being reached (ix and 16-17).

With regard to the quantitative dimension, Devlin (1965, 8) made it clear that the feelings of a bare majority of the population do not suffice to impose their views on the minority: "The fact that a majority of people may disapprove of a practice does not of itself make it a matter for society as a whole. Nine men out of ten may disapprove of what the tenth man is doing and still say that it is not their business." Rather, he referred to the standard of "the man in the jury box": "The moral judgement of society must be something about which any twelve men or women drawn at random might after discussion be expected to be unanimous" (15).

If we are to accept Devlin's view, then access to PGD and other ARTs could legitimately be curtailed not only when they are likely to cause actual harm but also when their use, or a particular use to which they can be put, provokes genuine and significant negative feelings in the "ordinary person." As Devlin made abundantly clear, the bare preferences of the majority do not provide a legitimate mandate for the state to intrude in otherwise private decisions.

Of course, this may still fall some way short of Dworkin's requirements for a genuine moral view. But, for present purposes, though their respective positions are frequently depicted as occupying opposite poles, one can note the agreement between Devlin and his liberal critics that mere popular opinion is not, in and of itself, sufficient to justify legal curtailment of individual liberty. Devlin (1965, 15) might have rejected the requirement that the behaviour in question be harmful (or prospectively so), but he was equally clear in his rejection of simple majoritarianism: "How is the lawmaker to ascertain the moral judgements of society? It is surely not enough that they should be reached by the opinion of the majority."

How does each of these approaches relate to decision making about ARTs? In particular, given the subject matter of this collection, what do they have to say about public participation in such decisions? Evidently, Devlin's approach would require that efforts are made to ascertain popular opinion – not only the number who agree with or support a particular mode of behaviour but also the intensity with which they hold their views. Do they harbour feelings of genuine disgust or indignation? Or are their views closer to mere dislike or disapproval? Clearly, to satisfy Devlin's requirements, any survey of public opinion must be sufficiently nuanced to gauge such differences.

For those more sympathetic to the liberal position, it might seem that popular sentiment has a less obvious role to play; in fact, a great deal of the concern of Mill, Hart, Dworkin, and their successors lay in protecting minorities from their disapproving neighbours. Yet, the determination of whether harms are likely to exist, and their probable intensity, is likely to be conducted more accurately if informed by wide-ranging consultation. The alternative, whereby a regulatory body such as the United Kingdom's Human Fertilisation and Embryology Authority (HFEA) (or, indeed, the government) attempts to predict which interests will be impacted by a particular policy without speaking to those likely to be most affected, seems to require an almost preternatural display of empathy.

A recent example, again from the United Kingdom, of such a futile attempt was the series of debates prior to the introduction of the Human Fertilisation and Embryology Act 2008. Section 14(4)(9) of the act requires that

> Persons or embryos that are known to have a gene, chromosome or mitochondrion abnormality involving a significant risk that a person with the abnormality will have or develop (a) a serious physical or mental disability, (b) a serious illness, or (c) any other serious medical condition, must not be preferred to those that are not known to have such an abnormality.

In proposing, and eventually introducing, this requirement, the government's primary objective – as set out in the explanatory notes to an earlier version of the bill[1] - seems to have been to preclude the sort of choices made by Sharon Duchesneau and Candy McCulloch, a US couple who selected a sperm donor who would maximize the chances that their future offspring would share their deafness (Spriggs 2002).

The ensuing backlash against the clause – spearheaded by the ad hoc, and now disbanded, StopEugenics group – was ferocious and seems to have caught the bill's drafters, and the UK government, completely unaware.[2] It also revealed genuine offence among substantial numbers of deaf people. As one petition expressed it, "Clause 14(4)(9) creates a situation whereby, in law, the life of a Deaf person becomes of lesser worth than that of a hearing person, despite the Government's aim for a more equal society."[3]

Such sentiments,[4] and the media attention that they received,[5] ultimately led to the removal of any reference to deafness in the final explanatory notes[6] (though the general principle is still contained in the legislation as enacted). It is possible, though, that the damage to relations with much of the Deaf community will last longer.

How much weight should be accorded to the offence evidently caused to some deaf people might divide adherents to the harm principle. The matter is further complicated by the inherent difficulty in attaching weight to the putative harm that the clause was intended to avoid (as I discuss later). It is also uncertain whether a more considered drafting of the explanatory notes, without specific reference to deafness, would have obviated much of that offence. However, the absence of prior consulation with deaf groups, activists, and individuals was a recurrent complaint from opponents of the clause. And a better understanding of such views would at least have better positioned the bill's drafters to weigh the various interests that Section 14(4)(9) would likely impact.

Sex Selection

If failure to consult at all can lead to problematic outcomes, with regulators and legislators failing to anticipate the nature or degree of interests involved, so can undue deference to public opinion. An example is the regulation of pre-implantation (and, as of 2009, pre-fertilization) sex selection in the United Kingdom. The Human Fertilisation and Embryology Act 2008 introduced for the first time an explicit statutory prohibition of "any practice designed to secure that any resulting child will be of one sex rather than the other" (Schedule 2, para. 3). Prior to this, the prohibition on the practice amounted to an HFEA policy that licences for sex selection would be granted only where the practice was employed to avoid the transmission of sex-linked genetic disorders.

In 2002-3, at the request of the UK government, the HFEA conducted a consultation on the regulation of sex selection. It consisted of several elements:

- research on social and ethical issues and scientific and technical issues;
- qualitative research conducted through discussion groups to investigate how individual members of the public approach and grapple with issues surrounding sex selection, what they see as being the central issues, and the concepts and language that they use to discuss them;
- consultation, informed by the concerns identified in the qualitative research, to invite more detailed statements of views from stakeholders and the general public; and
- a survey of a representative sample of the UK population, focusing on public perception of, and opinions about, the issues of greatest concern identified through the qualitative research. (HFEA 2003)

The results of both qualitative and quantitative aspects of the research revealed a substantial majority of respondents to be ill disposed toward sex selection for non-medical reasons: almost 83 percent of respondents to the consultation questionnaire (HFEA 2003, para. 91) and 67 percent of the opinion poll sample (HFEA 2003, Appendix F) disagreed with permitting PGD for non-medical sex selection. These findings seem to have impacted significantly on the Authority's ultimate recommendations: "We have been particularly influenced by ... the quantitative strength of views from the representative sample polled by MORI and the force of opinions expressed by respondents to our consultation. These show that there is *widespread hostility to the use of sex selection for non-medical reasons.*" The report goes on to explain the manner in which these findings led to the Authority's conclusion: "By itself this finding is not decisive; the fact that a proposed policy is widely held to be unacceptable does not show that it is wrong. But *there would need to be substantial demonstrable benefits* of such a policy if the state were to challenge the public consensus on this issue" (HFEA 2003, para 147).

It is clear, then, that the HFEA has rejected the approach of Mill and his liberal successors. Indeed, it has reversed the onus of the presumption of liberty, instead opting for an approach that places the burden of justification firmly on those who would make unpopular decisions. Does it follow, though, that an approach more in keeping with Devlin's societal morality approach has been taken? Certainly, the HFEA's claim of "widespread hostility" to social sex selection seems more likely to satisfy Devlin's test. Closer examination of the MORI survey's findings, however, reveals this conclusion to rest on a questionable empirical basis.

Leaving aside concerns about the extent to which a relatively small study sample can credibly claim to be representative of societal opinion (Harris 2005, 286), we might wonder whether the HFEA has paid undue consideration to the mere *numbers* of views expressed while affording insufficient heed to the *intensity* with which those views were held. Only 43 percent of those surveyed in the opinion poll *strongly disagreed* with the statement that "sex selection through testing embryos created outside the body should be permitted for 'family balancing.'" Although a further 24 percent responded that they "tend to disagree," giving an overall two-thirds majority opposed to permitting the practice, only a narrow majority (52 percent) of the public sample believed the issue to be "important."

Certainly, the responses to the consultation more closely approximated such a consensus. But, as the report itself noted, "a great many respondents felt that sex selection was unqualifiedly wrong because it involved interference with divine will or with what they saw as the intrinsically virtuous course of Nature" (HFEA 2003, para. 64). And since it concluded "that public policy in this area should be founded on wider considerations than those deriving from a particular set of religious beliefs" (HFEA 2003, para 134), it is difficult to know what weight the Authority attached to such views.

With regard to the discussion groups, no figures are available, though we are assured that "most [participants] felt that neither PGD nor sperm sorting should be available for less serious medical reasons, or for social reasons" (HFEA 2002, para. 6.7). From a deliberative perspective, this is the only really interesting part of the consultation process, and it is the part that displays the most sophistication and ambivalence in the views expressed.

On the one hand, participants expressed concern that "trying to exert control over their child's sex, or looks, or intelligence would be to deny one of the core values of having children" (HFEA 2002, para. 6.4). On the other, considerable intuitive support was shown for the presumption of liberty and the harm principle: "Certainly, in citizenship terms, parents in particular felt that it was very difficult to argue the case against giving one couple their heart's desire when no-one was harmed – however uncomfortable it made them feel personally" (HFEA 2002, para. 6.4).

Whatever their "end" position, the impact of couching the debate in terms of citizenship, and thinking about the issue from this standpoint, meant respondents thought that they had to produce robust arguments against sex selection for social reasons. They thought that the onus was on them to justify their seeming intolerance, and they struggled to do so (HFEA 2002, para. 6.2).

If we discount (as did the HFEA) the write-in consultation, which seems (not uncommonly) to have been dominated by unrepresentative religious and "pro-life" views, then the only indication of the extent and degree of opposition to social sex selection comes from the MORI opinion poll. On that basis, with only a minority expressing strong opposition to allowing the practice, and almost half of the sample not considering it an issue of particular importance, we might legitimately wonder whether Devlin's standard of injury against society as a whole has been satisfied. Certainly, when only 43 percent of the population are strongly opposed to social sex selection, it is not obvious that the ordinary person has reached the limits of tolerance.

Most clearly, such evidence does not seem to substantiate the HFEA's claim that "the great majority of the public are strongly opposed" to social sex selection, while reference to a "public consensus on this issue" seems unduly optimistic (HFEA 2003, para. 147). In short, even if we reject the harm principle, and accept Devlin's justification for defending a common morality, it is far from clear that his criteria have been satisfied in terms of social sex selection.

With regard to the discussion groups, it seems that the public, when prompted to consider such matters, is torn between respect for a "realm of privacy" and a desire to protect the children who would result from such choices – with the latter concern seemingly (though we are shown no figures) weighing more heavily with most participants. Although the report on the discussion groups makes interesting reading, it is impossible to ascertain from it the extent or degree of opposition to non-medical sex selection and therefore impossible to ascertain whether Devlin's benchmark has been met.

Two potentially highly relevant points do emerge, however: first, that something like the harm principle exerts considerable influence over the UK public (at least if we assume the discussion group participants to be representative); second, that they were, on balance, persuaded that a sufficient risk of harm was posed by the practice of non-medical sex selection to agree (however tentatively) that the practice should not be permitted.

Ultimately, the Authority seems itself to have attached significant weight to this possibility. In a subsequent debate in the *Journal of Medical Ethics,* Tom Baldwin, then deputy chair of the HFEA, referred to the importance of harm in the Authority's deliberations: "It is here that most critics of unrestricted sex selection seek to make their case, arguing that where sex selection is undertaken without medical justification there is a likelihood of harm to children whose sex is thus determined" (Baldwin 2005, 289-90). Is it then possible that, while Devlin's standard for consensus-based legal intervention might not have been met, Hart's standard for harm-based intervention might have been?

The Harm Principle Re-Engaged?
Although adopting the language of "welfare" and "damage" rather than "harm," the HFEA's final recommendations on non-medical sex selection arguably would satisfy the criteria required by Hart and Mill for legal intervention. In the report's conclusions, the Authority stated this view:

> The most persuasive arguments for restricting access to sex selection technologies, besides the potential health risks involved, are related to the welfare of the children and families concerned. There was some considerable alarm among consultation respondents that children selected for their sex alone may be in some way psychologically damaged by the knowledge that they had been selected in this way as embryos. Some consultation respondents expressed concerns that such children would be treated prejudicially by their parents and that parents would try to mould them to fulfil their (the parents') expectations. (HFEA 2003, para. 139)

In slightly more esoteric language, Baldwin (2005, 290) expressed a similar sentiment:

> A common theme of many of the responses the HFEA received was that the relationship between parents and their children would be distorted if parents were able to determine the sex of their children. Sometimes the point was expressed as the claim that sex selection would threaten the child's "otherness," its independence, and this claim points to a concern that motivated much of the public opposition to sex selection.

For reasons that I (and others) have explored in depth elsewhere (Gavaghan 2007; Glannon 2001; Harris 2007), it is no easy thing to attribute harm when discussing reproductive technologies, for the simple (though often counterintuitive) reason that the genetic attributes with which such a child is born are preconditions of *that* child being born at all. If a couple were denied the opportunity to use PGD to determine the sex of their child, they could elect not to have any more children;[7] they could elect to conceive by "natural" means (if that option is open to them); or they could go ahead with IVF but without the option of sex selection.

In any of those events, the likelihood is negligible that they will have the same child, made from the same gametes, as they would were they allowed to use sex selection. That being so, for the child actually born as a result of sex selection, the only alternative to risking psychological harm later in life is never to be born. If this line of reasoning is accepted, then it follows that it will rarely, if ever, be coherent to claim that a child's birth should be prevented on the basis of concern for his or her welfare.

Hardly surprisingly, this conclusion has struck several observers as less than persuasive (Peters 1999). This is not the place to explore that debate in greater depth, but certainly, among bioethical commentators, there is nothing resembling a consensus on issues of welfare and harm regarding future lives and, in particular, whether it is even coherent to speak in terms of a child being harmed by an act or choice on which its existence was contingent. Indeed, this lack of consensus seems to have been reflected in the HFEA's own deliberations: "There was in fact no consensus among the members of the HFEA on this matter. Instead there was an agreement that this is an issue where there are powerful arguments on both sides of the case, but also that *neither side can seriously maintain that immediate serious harm is in prospect*" (Baldwin 2005, 290, my emphasis). The last clause of that paragraph is of particular interest, for it seems to be an acknowledgment, by the HFEA's deputy chair, that the principal condition for the harm principle was not satisfied – or at least that the members of the Authority could not agree that it had been.

Proponents of the Mill-Hart approach would regard that as reason enough to stick with the presumption of liberty and leave choices about sex determination to individuals or couples; as Harris writes, "If as Baldwin states, there are powerful arguments on both sides then the presumption must be in favour of liberty" (Harris 2005, 287; see also Tizzard 2004). In fact, the HFEA seems to have adopted a very different approach. In the absence of clear evidence of, or agreement about, the prospect of harm, it reverted to a majoritarian default, according to which, "with all the evidence described earlier which showed public opinion very largely on one side of the argument, the HFEA accepted that it was reasonable that public policy should be guided by this public consensus until the argument is further resolved" (Baldwin 2005, 290).

Yet, as I have argued, that weight of public opinion seems to fall considerably short of the conditions set by Devlin for legal intervention with personal choice in that – insofar as the HFEA's poll can be relied on – most people do not seem to view non-medical sex selection with genuine antipathy. If it profoundly offends some genuinely felt community value, it is unclear what it might be.

Conclusion

It is sometimes asserted that bioethicists and medical lawyers are in thrall to the notions of autonomy and liberty. In a recent polemic, barrister Charles Foster (2009, 182) described autonomy as "the sole language of academic discourse" and described the phenomenon of "a sort of dogmatic, hectoring liberalism replacing Christian morality as the Universal Law." Indeed, my own contributions to this debate have seen me diagnosed with "a severe dose of neoliberal ideological bias" (King 2007, 13).

I make no secret of, nor apologize for, my liberal (though not "neoliberal"!) instincts; I find the idea of a realm of private decision making, not subject to the "peer review" of our neighbours, appealing, as I suspect many people do. It is not, though, my purpose here to make the case for such an approach to PGD or sex selection in particular (though I have done so elsewhere; see Gavaghan 2007). Rather, it is to consider – from the perspective of someone whose professional life involves scrutinizing and critiquing the law – the various options available to lawmakers with regard to how public views should inform their task.

If I am to avoid begging too many questions, the first possibility that I should perhaps confront is whether there is a case for ARTs to be regulated on a simple majoritarian model. Slovenia did something like this in 2002 when it held a referendum on access to assisted reproductive technologies. Such an approach has the advantage of being well insulated from allegations of elitism. Yet, there is no way to guard against (and often every reason to suspect the existence of) the sort of ill-informed and ill-considered prejudice that so concerned Ronald Dworkin. Furthermore, such a model seems to obviate much of the current role of the HFEA; rather than establishing a body to evaluate evidence and argument, the relevant government ministry could simply employ researchers to sample public opinion or – in the unlikely event that money was no object – hold a full-scale referendum, with policy based on whatever the majority decided.

On the other end of the liberal-populist spectrum, strict adherence to the harm principle seems likely to lead to a decidedly laissez-faire approach to sex selection and to the use of ARTs more generally – at least in the absence of any evidence that demographic distortion is likely, as the HFEA itself conceded, or until a credible answer is offered to the observation that almost *no* children can be said to be harmed by being born.

In the middle lies a range of approaches that seeks to attach some measure of significance to views that are widely held within a society but that require more than a simple show of hands. Devlin would require that the ordinary man or woman in the jury box experience sincere feelings of disgust, whereas Dworkin would require that the views in question are considered, consistent, and at least minimally informed.

I have argued that the data on which the HFEA relied in framing its policy did not, in fact, demonstrate either that a majority within UK society felt strongly averse to sex selection or that their views were particularly considered. However, I do not seek to exclude the possibility that important and widely shared moral values *are* offended by the practice. The research conducted by Scully, Shakespeare, and Banks (2006, 753), for example, found a strong element of support for "the idea that children should be a 'gift' (and not a 'commodity' or, less frequently, a 'right')." They explained that the "gift" idea was almost always employed metaphorically – few participants expressed much sympathy for the theological notion of a divine "giver" – to convey the idea that children "should be accepted as they are." This idea certainly seems to have had an impact on at least some members of the HFEA (Baldwin 2005, 290) and bears substantial similarities to a view expressed – in more traditional academic form – by Jürgen Habermas (2003) and Michael Sandel (2007) – though the former's arguments have been dismissed, in typically robust fashion, by John Harris as "mystical sermonising" (Harris 2005, 286 and 288).

It is no criticism of the research of Scully and colleagues to say that, in itself, it forms an insufficient basis on which to form any sort of regulatory policy – they themselves explicitly recognize as much. For one thing, any attempt to ascertain whether such a concern is widely held within UK society would require a substantially larger sample size than the forty-eight participants in their study groups. For another thing, there would have to be some attempt to gauge the intensity with which such views are held. As with the HFEA's discussion groups, a degree of ambivalence seems to exist between moral disapproval of social sex selection and a reluctance to impose that disapproval on others; as one participant put it, "I would find it very difficult to impose that principle on somebody else, because I wouldn't know what their situation was ... I wouldn't be able to say no, you shouldn't do that (Social worker)" (Scully, Shakespeare, and Banks 2006, 759). (The authors go on to note, however, that – notwithstanding this reluctance – "knowing that this line was already drawn [since UK law currently forbids SSS] meant they saw no reason to change it" [2006, 759].) This is interesting. Is it perhaps a case of emotional/psychological inertia, where any departure from the status quo faces the burden of justification? Or were participants reluctant to impose such a decision themselves but happy enough to learn that someone else had done so?)

Determining whether this concern about excessive parental control is indeed one of our society's core moral values might also require seeking to ascertain whether it manifests itself with any degree of consistency in other areas of life. We might, for instance, expect to see similar unease when parents try to shape their children's lives in other ways – arranged marriages, denominational schooling, and "the low-tech, high-pressure child-rearing practices we commonly accept" (Sandel 2007, 60-61); all seem like plausible candidates for similar concerns. If the imposition of gender stereotypes is really a concern, then this might be evident in support for the likes of the recent PinkStinks campaign in the United Kingdom, whose website bears the following mission statement: "We believe that body image obsession is starting younger and younger, and that the seeds are sown during the pink stage, as young girls are taught the boundaries within which they will grow up, as well as narrow and damaging messages about what it is to be a girl" (www.pinkstinks.co.uk).

Of course, these other issues and campaigns can differ from social sex selection in various relevant ways. My point, however, is that we should expect the putative aversion to excessive parental control, or respect for the notion of "givenness," to be manifest in other places; otherwise, we should be skeptical of the claim that it represents a widely and strongly shared moral value as opposed to a mere knee-jerk aversion to "the new."

If we are to follow Devlin in relying on popular morality to justify legal coercion of dissenting minorities, we must first satisfy ourselves that the morality in question is genuinely popular, that it is shared widely, and that adherence to it is widely thought to matter. This, I suggest, involves asking people not merely whether they approve of a practice but also whether they are sufficiently resolute in their opposition to support its criminalization. It is possible, after all, to express disapproval of all manners of behaviour that we would not want the law to punish; as Hart (1963, 76) once wrote, "it is a disastrous misunderstanding of morality to think that where we cannot use coercion in its support we must be silent and indifferent." It might be that, confronted with the question in that manner, support for a "realm of private decision making," where the law does not intrude, is as much a part of our societal morality as are particular views about reproductive technologies. Or it might not. But unless and until the question is asked in that way, we will not know.

Which brings me to a final observation. Is it possible that, in asking people for their opinions about the morality of certain choices, researchers almost beg the question of whether it is their business at all? No government agency would, presumably, think of commissioning official research (as opposed to surreptitious focus groups) to ask whether it is appropriate for the prime minister to be married to a Roman Catholic (as Tony Blair was) or for a cabinet minister (Chris Smith) to be homosexual; these are properly regarded as private matters and – bluntly – no one else's business.

Although it does not, of course, follow that the same should be said of social sex selection, it is perhaps worth considering whether the very choice of which topics about which to consult sends a message to those consulted to the effect that – almost by definition – the subject in question does not properly lie within the realm of private morality.

NOTES

1 Bill introduced in the House of Lords on 8 November 2007, paragraph 109, http://www.publications.parliament.uk/.

2 This is based in part on my personal observations at a meeting with members of StopEugenics and the UK Department of Health in April 2008. For a summary of that meeting, see http://www.bionews.org.uk/.
3 See http://www.grumpyoldeafies.com/.
4 See also Atkinson (2008); Bradshaw (2008); Emery et al. (2008).
5 See, for example, "Deaf Demand Right to Designer Deaf Children," *Sunday Times,* 23 December 2007.
6 Paragraph 114 of the final explanatory notes reads thus: "Embryos that are known to have an abnormality (including a gender-related abnormality) are not to be preferred to embryos not known to have such an abnormality. The same restriction is also applied to the selection of persons as gamete or embryo donors. This would prevent assisted reproduction technology being used to select an embryo with a view to increasing the chance of giving birth to a child that had or would develop a serious medical condition, or to select a donor to increase the chance of a child having a serious medical condition." Hence, no specific examples of what might be regarded as a "serious medical condition" are identified.
7 This seems likely to have been true of Alan and Louise Masterton, the Scottish couple who wanted to use the technology to "restore the female dimension" to their family after the death of their only daughter.

REFERENCES

Atkinson, R. 2008. "I Wouldn't Have Minded if My Baby Had Been Born Deaf, but the Embryology Bill Suggests I Should." *Guardian,* 10 October.

Baldwin, T. 2005. "Reproductive Liberty and Elitist Contempt: Reply to John Harris." *Journal of Medical Ethics* 31: 288-90.

Beyleveld, D., and R. Brownsword. 2001. *Human Dignity in Bioethics and Biolaw.* Oxford: Oxford University Press.

Black, J. 2001. "Decentring Regulation: Understanding the Role of Regulation and Self-Regulation in a 'Post-Regulatory' World." *Current Legal Problems* 54: 103-47.

Bradshaw, H. 2008. "Why It Should Not Be Illegal to Implant 'Abnormal' Embryos." 21 January. http://www.bionews.org.uk/.

Brownsword, R. 2008. *Rights, Regulation, and the Technological Revolution.* Oxford: Oxford University Press.

–. 2009. "Human Dignity, Ethical Pluralism, and the Regulation of Modern Biotechnologies." In *New Technologies and Human Rights,* edited by T. Murphy, 19-84. Oxford: Oxford University Press.

Cane, P. 2006. "Taking Law Seriously: Starting Points of the Hart/Devlin Debate." *Journal of Ethics* 10: 21-51.

Committee on Homosexual Offenses and Prostitution. 1957. *Report of the Committee on Homosexual Offenses and Prostitution.* London: Her Majesty's Stationery Office.

Devlin, P. 1965. *The Enforcement of Morals.* Oxford: Oxford University Press.

Dworkin, R. 1966. "Lord Devlin and the Enforcement of Morals." *Yale Law Journal* 75: 986-1005.

Emery, S., T. Blankmeyer Burke, A. Middleton, R. Belk, and G. Turner. 2008. "Clause 14(4)(9) of Embryo Bill Should Be Amended or Deleted." *British Medical Journal* 336 (7651): 976.

Foster, C. 2009. *Choosing Life, Choosing Death: The Tyranny of Autonomy in Medical Ethics and Law.* Oxford: Hart.

Gavaghan, C. 2007. *Defending the Genetic Supermarket: The Law and Ethics of Selecting the Next Generation.* London: Routledge-Cavendish.

Glannon, W. 2001. *Genes and Future People.* Westview Press.

Habermas, J. 2003. *The Future of Human Nature.* Polity Press.

Harris, J. 2005. "No Sex Selection Please, We're British." *Journal of Medical Ethics* 31: 286-88.

–. 2007. *Enhancing Evolution: The Ethical Case for Making Better People.* Princeton: Princeton University Press.

Hart, H.L.A. 1963. *Law, Liberty, and Morality.* Oxford: Oxford University Press.

Human Fertilisation and Embryology Authority (HFEA). 2002. "Sex Selection: Policy and Regulatory Review – A Report on the Key Findings from a Qualitative Research Study." London: HFEA.

–. 2003. *Sex Selection: Options for Regulation.* London: HFEA.

King, D. 2007. "Preimplantation Genetic Diagnosis and 'Slippery Slopes.'" *Bionews* 407: 13.

Mill, J.S. [1859] 1979. "On Liberty." In *Utilitarianism,* edited by M. Warnock, 126-250. Glasgow: Collins Fount Paperbacks.

Peters, P.J. 1999. "Harming Future Persons: Obligations to the Children of Reproductive Technology." *Southern California Interdisciplinary Law Journal* 8: 375-400.

Sandel, M.J. 2007. *The Case against Perfection.* Cambridge, MA: Harvard University Press, Belknap Press.

Scully, J.L., T. Shakespeare, and S. Banks. 2006. "Gift Not Commodity? Lay People Deliberating Social Sex Selection." *Sociology of Health and Illness* 28, 6: 749-67.

Spriggs, M. 2002. "Lesbian Couple Creates a Child Who Is Deaf Like Them." *Journal of Medical Ethics* 28: 283.

Tizzard, J. 2004. "Sex Selection, Child Welfare, and Risk: A Critique of the HFEA's Recommendations on Sex Selection." *Health Care Analysis* 12, 1: 61-68.

Wolfenden Committee. 1957. *Report of the Committee on Homosexual Offences and Prostitution.* London: Her Majesty's Stationery Office.

Challenges to Deliberations on Genomics

MICHIEL KORTHALS

In many European countries, consultations on new technologies and other types of innovation are established. In the Netherlands in the 1970s, a large national debate was undertaken on nuclear energy and later, in the 1990s, on genetic modification. For decades, citizen juries have been organized in Denmark. England experimented with consultations in GM Nation. This participatory turn has been accompanied by intensive theoretical debate about the function and structure of deliberations as a complement or alternative to normal democratic procedures (Dryzek 1990, 2000; Gastil and Levine 2005; Mouffe 1996; Renn and Webler 1996). In 1971, Habermas published a short and obscure article in which he explained, for the first time, that discourses are the main vehicles for reaching the truth of propositions (cognitive discourses) and the justice of ethical judgments (ethical discourses). It took years before he, social philosophers, and social scientists developed this idea into more practical political and ethical organizational forms, and in the 1980s, the first publications appeared that followed this idea more broadly, and the new democratic perspective, coined deliberative democratic, emerged (Einsiedel and Eastlick 2000; Renn and Webler 1996; Schot 2001).

The history of this type of democracy is therefore not very old, only forty years. However, in those years, several versions of deliberative democracy have been developed, implemented, and critically assessed. My philosophical involvement with this type of democracy regarding science and technology started with the article of Habermas, and now, more than forty years later, I am still busy trying to make sense of this idea and give it meaning in regard to coping with social and technological developments. Currently, the deliberative approach is plural, often plagued by severe misunderstandings, and characterized by impressive conceptual and empirical progress and problems (Hamlett 2003).

In this chapter, I first give a short sketch of several versions of the deliberative approach and my version of it. Then I discuss a few problems that emerged during my work as an ethicist of food and agriculture. They include problems of unequal participation in deliberations and cognitive and normative uncertainties that pervade the life sciences, such as biotechnology, sciences that comprise different scripts (e.g., nutrigenomics: personalized or public health), and the framing of problems. Next I discuss multi-level problems, in particular, the relationships among local, regional, national, and international opinion formation and decision making. I then turn the table and argue that some scientific and technological projects have large and severe deliberative impacts while others do not. Discussing these challenges, I develop tools that include mapping different arrangements of the interaction between genomics and societal developments; developing imaginary futures through aesthetic explorations and scenarios, different moral screenplays, and dramatic rehearsals; fostering deliberative leadership; and designing deliberation-eliciting technologies.

What Is Deliberation?

There are many types of deliberation (Rowe and Frewer 2004; Rowe, Marsh, and Frewer 2004) and many theoretical perspectives on deliberation. I start with the proposal of Habermas (1971). According to him, deliberations are types of communication somewhere in between the three large social categories of government (administration), market, and civil society. He distinguishes three kinds of discourses: cognitive, about truth claims of descriptive statements; normative, about claims of the rightness of ethical judgments; and expressive, about claims of the authenticity of emotional expressions. In *Between Facts and Norms: Contributions to a Discourse Theory of Law and Democracy* (1996), Habermas designs a deliberative process model of democratic politics. This model features a multiplicity of mutually entwined forms of communication. The three types of discourse are supplemented in this model by strategic negotiations aimed at fair compromise and supplemented by legal discussions that evaluate the results for legal consistency.

The results of these discourses are principles universally agreed on; however, he concedes that, in the normative discourses, it is never merely universal norms that are produced; time- and place-related values and norms of communities are hidden in universal norms.

Habermas strongly believes in the necessity of separating universal moral norms (in rational, i.e., moral discourses) from the values of a community that he defines as "particular"; this distinction is strict and difficult to reconcile with his concession that universal norms are saturated with particular values. Moreover, he does not concede vice versa that particular values can have universal moral meanings: for instance, the food choices of people (expressing their values) can have universal moral meanings by incorporating (or not) universalizable values such as biodiversity, sustainability, and respect for the good life of future generations. Food choices that people make (be they producers, consumers, or, in facilitating them, governmental bodies), such as consuming veal tongue from calves raised in isolated crates, hamburgers and other fast food, or cod, have direct ethical impacts on animal welfare, the environment, and biodiversity. Many have argued against this distinction between moral norms and particular values (Benhabib 1996). Habermas's restriction of ethics to moral, universal principles implies that ethics cannot take into account the grey zone between universal norms and universal aspects of the good life and cannot assess the ethical meaning of the particular good life to which people aspire.

Related to this is Habermas's presupposition that, with respect to discourses on universal norms, one should always strive for a moral consensus and produce only one right solution, as can be seen from the following statement: "The democratic process promises to deliver an imperfect but pure procedural rationality only on the premise that the participants consider it possible, in principle, to reach exactly one right answer for questions of justice" (Habermas 1998, 403).

Regarding this second point of criticism, that moral decisions do not allow for a variety of good solutions, in other publications Habermas has tried to make it clear that the binary feature of moral judgments (either right or not right) belongs exactly to their cognitive (rational) meanings. He presupposes, for instance, that decisive reasons on moral judgments on the moral meanings of food always can and should be given. Habermas makes it clear that rightness is only connected to normative judgments. With the validity claim of rightness connected to a normative judgment, the speaker asks for agreement, and, if his opponent does not agree, evidence has to be given that, indeed, the normative judgment is right. Consequently, Habermas calls discourses and outcomes that are the result of compromise, fair negotiation, or prudent consideration not really moral and rational.

Other versions of the deliberative model, like that of Benhabib (1996) or Bohman (2007), have criticized the unnecessary limits that Habermas puts on the potentialities of discourse (or deliberation) and reasoning and his emphasis on strictly universal discourses. Bohman points out that, even in the moral type of discourse, often not decisive but plausible reasons play a dominant role. The continuous emergence and uncertain implications of new scientific and technological developments in the field of food and nutrition make it nearly impossible for only one answer to be the right one (see below). It seems to me that Habermas's perspective insufficiently takes into account diverging worldviews, their concomitant controversies and dilemmas, their ethical impacts, and the potentialities of ethical and moral deliberations about them.

Finally, Habermas excludes with this intrinsic connection between rightness and individual judgments ideas of the good life from the moral domain; these later ones are mostly shaped in the form of narratives and cannot, by implication, be rationally discussed. Testing a rightness claim means scrutinizing the arguments for or against a certain ethical judgment. However, most individual ethical judgments get their meanings from those narratives, as do other cognitive judgments, and they function within a larger framework or vocabulary that, in the case of severe doubt, also has to be discussed, which is impossible, however, in a rational discourse à la Habermas.

Thus, in my view, rational deliberation does not always require consensus concerning moral problems regarding food. One should make a distinction between deliberations, aiming at the development of interesting opinions and narratives, where no consensus is necessary (see types c and d in Table 6.1), and (rational) deliberations, aiming at individual decision making that might or might not be based on consensus (types a and b). In a fully fledged "pragmatist ethics," these four types of deliberation should be taken into account.

In this matrix are four possibilities for deliberations, and it is clear that Habermas, in his deliberative approach, only gives attention to the first two. However, types c and d cover the framing of ethical issues, which mostly happens from a dominant or made-dominant vision of the good life, no matter how rarely this is made explicit. Finally, the distinction between the four possibilities also makes it clear that deliberations never meet fully

TABLE 6.1

Possibilities for deliberations

Tasks for a pragmatist ethics	Product	Process
Rationalist ("context of justification")	(a) *Traditional ethics* Providing arguments and justifications for or against courses of action	(b) *Discourse ethics* Structuring and safeguarding fair public deliberation and decision making
Romantic ("context of discovery")	(c) *Dramatic rehearsal* Criticizing and renewing vocabularies, exploring possible future worlds	(d) *Conflict management* Aiding an open confrontation of heterogeneous moral vocabularies and worldviews

Source: Keulartz et al. (2002, 203).

democratic requirements of parliamentary democracy and, therefore, never can be an alternative to this strong (by the vote of all) legitimized type of decision making. Deliberation in the sense of type a is part of parliamentary debates and decision making, and, therefore, deliberations are always complementary to normal democratic processes. Deliberations (b-d) are more like the Socratic dialogues about intriguing problems of life in the market (*agora*) in Athens than like the discussions and decision-making procedures in the Athenian republic.

Why Deliberations?

Although deliberations are often used in consulting the public about technological innovations, the reasons to do so can vary (Burgess 2003; Einsiedel and Eastlick 2000). Consultations and deliberations are organized everywhere in Europe (e.g., Understanding Risk Team 2004). According to the EU General Food Law, Articles 7, 8, and 9, there is a need for consultations in the case of innovations.

I can discern at least five reasons that the public has a right to ask for deliberations about science and technology innovations and what motivates governments to organize them. The first reason is purely strategic: because the scientific and technological applications will be used by members of the public, they should get information on an innovation. Often this reason is used by scientists, who say that science needs the support of society, and that is why they ask the public to become involved. If the public refuses to use the innovation, it is over, whatever the reason might be. So, purely based on instrumental reason, it is necessary for science to know as far as possible in advance what the public wants and does not want. This is a reason for a fairly restricted way of communicating: only provide information that influences the public in favour of the desired use, and abstain from talking about risks and losses. A complicating factor is that the science community is not a political and social unity, so some scientists will give more and different information than others, for social or strategic reasons. I guess that science in general will not succeed in offering unifying, restricted information in this sense.

The second reason is that, because the public pays taxes, it pays for the science and technological projects, so it has grounds to participate in decision making. In this case, science has some motivation not only to give positive advice but also to do its best in providing information on the risks. Although politicians ultimately will decide, the public has some motive to become involved. A variant of this argument could be that, owing to the normative implications of science, the public should have some say in it, because, with respect to normative reasoning, scientists are not the only experts. This can be called the normative or democratic argument.

A third reason to involve the public is epistemological. It is often said that experts incorporate only a limited type of knowledge and that the whole, the combination of this detailed knowledge in a higher way, is up to the public. In this case, the public is said to function as a source of knowledge, not only as a source of values, as in the previous argument. Some even say that laypersons can be experts on certain aspects, such as identifying risks (see above) and taking measurements in the case of ecological characteristics. For example, data on bird migration are mostly collected by laypersons, and other examples can be given of the knowledgeability of so-called laypersons. However, in general, I think that these claims of epistemological symmetry of laypersons and science experts are exaggerated, though in some individual instances it can be the case. So we should be careful not to paint the public as not knowledgeable, but, in general, science seems to me to be a more methodologically advanced form of knowledge production.

A fourth reason is that innovations have consequences with which end users and other involved persons will have to cope. Even people who are not taxpayers or otherwise involved can be affected by an innovation. When I decide to undergo DNA screening, the information can have dramatic consequences for my relatives. When a new, higher-yielding crop is used, more water will be used, so non-agricultural people can be affected, to name one effect.

The fifth reason is, in my view, the most compelling, because it takes into account that values do determine the considerations of scientists in general and in particular with respect to their decisions on significant truth and consequently on research priorities. Why is so much research done on sexy topics and not on others (e.g., chronic diseases)? A very moderate and sophisticated philosopher of science, Kitcher (2002), in *Science, Truth, and Democracy,* presents the view that science is not about truth but about significant truth: significance is reached by the choices made to do research on one topic instead of another. As Kitcher makes clear, it is here that science should listen to the values, reasoning, and standpoints of the public. Moreover, the exchange of opinions on important innovative issues gives citizens the possibility of listening to each other and of understanding their sometimes deep differences. This deliberative point of view, from my perspective, is the most compelling argument for engaging the public in deliberations on research priorities (and only if necessary on the data or theories).

When Are Deliberations Fruitful?

Citizens of modern democracies have busy lives, and it seems to me exaggerated to require them to participate continuously in all kinds of deliberations. Their interests and values differ, and so do their motivations to participate positively. Moreover, deliberations do not have to be organized in all situations, such as in acute emergencies or (the reverse) circumstances in which routines such as democratic procedures are sufficient. The functions of deliberation are, as has been discussed earlier, to make possible exchanges between different opinions on important innovations and to look for new ways of handling both the differences in opinion and the innovations. So what are their functions? Mostly, the function of debates is not decision making but enhancing the insight into cognitive and normative issues connected with science. Consultations are not a panacea or easy instrument. They are not always necessary – nobody is saying this. The structure and main issues are dependent on the aim. In my view, the aim should be the process of increasing the awareness of participants of the ethical and social problems of new sciences and not always decisions or regulations with respect to these sciences. This process will probably have very divergent outcomes, but parts of these outcomes can be used in the ensuing regulatory and science policy-making processes.

In the discussion of the human genome, for example, it became clear from the ethical debates that the public was seriously worried about issues of privacy in regard to the new genomic diagnostics in health. What would happen if an employer or bank knew that it had certain vulnerabilities? On the basis of these concerns, the genomic scientists fundamentally changed their research and responded in a way that made this concern less worrying.

If process is indeed the most important factor, then one of the main issues of organizing a debate is who is represented. One should reflect on the role of NGOs: they should not be the only spokespersons, but, let us admit, they are also not exclusively power-driven or self-interested groups. NGOs are mostly idealistic but one-sided groups. Some have an argument-based style; some are much more dogmatic and deeply entrenched in their views and not accustomed to listening to arguments. It is an exaggerated description of their power and agency, however, to say that they invented social unrest. We can only argue that Greenpeace can socially amplify the perception of certain risks because of a broader public lack of trust. However, the main issue remains, whom do they represent?

What Is Successful Deliberation?

There is in the literature considerable difference of opinion regarding the successful effects of deliberation. For instance, Chambers (2003, 318) argues that "a central tenet of all deliberative theory is that deliberations can change minds and transform opinions." Debates are said to bring people in contact with each other and their opinions, which implies that they change their opinions. However, changing an opinion is not the most important criterion of the success of deliberation. People often change their opinions due to all kinds of factors. More important is what their motives are in changing their opinions. In the case of participating in a deliberation, people can stick to their original opinions but on the basis of different arguments and/or motivations. Was it a new insight or some new way of seeing things? Did they broaden their perspectives so that they could provide better arguments for their opinions? And, when changing their opinions, did they do so because of the presence of a large majority and because cognitive disagreements seemed fruitless or because the majority was indeed right?

Other important questions that determine the success of deliberations are what will be done with the output of the deliberations? Who participates, and who represents which party or part of the public? What will be on the agenda? Will the issues that interest or concern the public be talked about?

Five Challenges to Deliberations

Unequal Participation

Not everyone wants to deliberate all the time. Many people like a kind of "stealth democracy": they are satisfied with the normal parliamentary decision making that, according to their view, goes smoothly and does not bother them (Theiss-Morse 2002; Theiss-Morse and Hibbing 2005). Other people do not engage in deliberations because they fear social ostracism or expect that they will not be heard. This is a relevant concern because, without the inclusion of these people, deliberations run the risk of being a kind of "enclave of gated democracy," a play toy for the well educated (Feenberg 2001). Inequalities, and the unwillingness of dominant parties to deliberate transparently, deter people from participating in debates. The contexts of institutionalization of deliberations and, in particular, power relations are therefore points of concern. However, people have opinions and are usually willing to share them and debate them in cases in which they are really involved (Delli Carpini, Cook, and Jacobs 2004). Moreover, as Delli Carpini and colleagues report, there are many good studies showing that people in daily talk indeed construct and develop a kind of ethical discourse. It is not that people do not want to discuss ethical issues. Depending on the motivations, issues, expectations, and social contexts of power, they are willing to participate in more formal settings, for instance concerning a certain technology.

The construction and development of deliberations in these complex societal and technological settings must therefore be, first, as transparent and inclusive as possible. The agenda must be clear and should not be changed because some dominant player wants it changed. Second, with respect to the process, the deliberations about humans and material circumstances must proceed with expert input that is as impartial as possible. During the process of deliberations, not only talk but also extra-verbal mechanisms can be applied to try to deconstruct fixed positions and tense relations and bring to understanding seemingly strange orientations. Play and aesthetic performances can open up possibilities and make entrenched positions fluid, with the result that the deliberations are not repetitions of already well-known moves. Third, the output of the deliberations should have an effect on decision making. Fourth, deliberations are components of a learning process, and some kind of public awareness of their main outcomes needs to be kept lively, for instance in media attention (Delli Carpini, Cook, and Jacobs 2004; Korthals 2008a, 2008b).

Fundamental Uncertainties of Science and Technology

The second challenge to deliberations on scientific and technological issues is that they are often characterized by dynamic and uncertain developments. In the 1950s, would one have organized deliberations on the pill that is now called the "anti-baby pill"? The theme of the debate then would have been the healthy regulation of the female period (Keulartz et al. 2002). It took some time before the final function of this pill was discovered. When deliberations are organized around the ambitions and expectations of scientists who are directly involved, one runs the risk of the deliberations missing the point. Development of the science presents a dynamic in which "organized skepticism" (Merton 1968) follows organized utopianism: taken-for-granted truths and certainties are destroyed or, better, unmasked as untruths and uncertainties. The persistent intellectual dispute among competing research teams, in which established truths fall prey to critical scrutiny, is an indication of the mature evolution of the scientific understanding of the world. Scientists claim the right to make both promises and mistakes; that is the core of the scientific ethos (Kuhn 1962; Merton 1968). This oscillating history shows that, during the period of emerging paradigms or disciplines, scientists are in need of an organizing utopian idea that rallies their energy and directs their attention. Organized utopianism (promises) seems to be necessary too and will give occasion to mistakes and false promises; organized skepticism will unmask them later. The history of science is full of these (later discovered) mistakes, which, in some cases, can turn out to be big problems or even catastrophes. Examples of serious mistakes made in agriculture and food sciences are the use of lead, radioactivity, and DDT and the claim that all vegetable oils are healthy (Bryson 2003, Chapters 7, 10).

The development of innovations in nutrigenomics is a case in point because its short history has had expectations (personalized nutrition and the individual gene passport) but now faces cognitive and normative uncertainties galore (e.g., multiple functions of genes; see Pearson 2006; Piatigorsky 2007; Sriram et al. 2005). Often, more knowledge does not reduce fundamental uncertainties in research, innovation, and application, and a fixed state of technology is often an illusion.

Deliberations can take this challenge into account by paying attention to the process of producing and managing scientific uncertainties. Deliberative strategies to tackle uncertainties can involve separating plausible from implausible uncertainties and making a hierarchy; creating scenarios (Peterson, Cumming, and Carpenter 2003), stories, dramatic rehearsals on future strategies, and chains of events; and constructing emotional strategies for living happily with uncertainties (Korthals and Komduur 2010).

Framing the Problem and Inscribing the Script

The selection and development of scientific and technological innovations are steered by networks of researchers, policy makers, and material things and their concomitant values. These factors "inscribe," in the words of Akrich (1992, 208), how the technology will look. For instance, these factors delegate responsibility to certain actors and not to others and give occasion to different moral judgments and behaviours of actors midstream and at the end of the developmental path, which Akrich calls a script. Others, starting with Goffman (1974), call this process of inscribing a script "framing." It makes quite a difference when one frames food and nutrition as a public health issue, a personal health issue (Komduur, Korthals, and te Molder 2008), or a cultural issue or when one frames medicines in terms of the needs of an inhabitant of the West or of those of the South. When the script or frame of a technology is successful, societal groups align with the technology; if not, then the technology will not succeed.

Deliberations that take technologies at face value, and do not question the problems that they are supposed to solve and their prioritization, will fail in the end because they take the wrong definition of the problem as a starting point. For instance, when malnutrition is framed as a medical problem (and food is framed as a health issue), only medical and not agricultural strategies will be seen as solutions. A different definition of the problem can change the range of solutions seen to be plausible.

An important item of deliberation should therefore be the analysis of the given framings of a problem and alternatives. This attention to definitions of problems makes deliberations different from stakeholder analysis because the latter approach takes for granted the problem defined by the parties involved and the way that those parties define themselves (Hamlett 2003).

Attention to framings and to how participants define their roles often requires unconventional strategies that transcend mere talk, such as aesthetic explorations (e.g., plays), role changing, creating scenarios, and challenging dystopian or utopian proposals for technological innovation (Komduur, Korthals, and te Molder 2008).

Multi-Level Problems

A fourth challenge has less to do with science and technology and more to do with the increasing complexities of the social networks in which they develop. Modern states are subject to "territorial dispersal" in the sense that many important decision-making procedures are delegated to either supra-national levels or subnational levels (Scharpf 1999). Moreover, "horizontal dispersal" means that non-state actors contribute to policies (Hajer and Wagenaar 2003). The multi-level governance of innovations puts delibera-tions with their often local orientations at a disadvantage when they are not structured in a translocal way. Debates about genetic resources, genetic modification of crops or animals, or research priorities or patents have such a widespread, even global, impact that the local deliberation should be connected with vertical levels (Goodin and Dryzek 2006). An important principle here is that of subsidiarity, which puts responsibility and decision-making power in the localities and on translocal levels only when it is abso-lutely necessary. Moreover, in cases in which it is necessary to connect the local with translocal levels, it is advisable to put more emphasis on delibera-tive leadership. Accountable spokespersons of (local) deliberations can bridge differences among the levels involved. Another device can be to con-struct bridge organizations among the various deliberations (Keulartz and Leistra 2008).

Different Deliberative Impacts of Technologies

Deliberations are mostly started when technological innovations are already under way; even upstream consultations, taking place as early as possible, are a response to certain technological ideas. However, instead of waiting for societal reactions, a deliberative approach could try to stimulate early ideas on innovation that stir up societal debate. Some technological ideas are more controversial than others and can give rise to new and fruitful moral vocabularies. In the beginning of the GM debate about animals, a new ethical vocabulary emerged, such as that of animal integrity; nowadays this debate could use some new input to put more pressure on people to find new ways of thinking and handling these types of biotechnologies. Current-ly, the ideas on in vitro meat can function as important incentives for debate and for new views on eating conventional meat, which now can be seen as eating a corpse. Technological and scientific projects, even in their immature phases, can start new debates and, in particular, avoid the stalemates that often bring debates to a fruitless and repetitive end. The first technological reaction to the scarcity of mineral fuels and climate change, in the form of biofuels, inspired a debate about biofuels in which it became clear that land, water, and other resources are important for food and even more so for meat production. One can call these technologies "world disclosing" because they inspire one to look differently at the world. Compared with other technologies that do not stimulate debate, such as that of mobile phones, these disclosing technologies function well in enhancing moral debates (Driessen and Korthals, forthcoming).

Next to art and social movements (other moral-disclosing agents), one could try to design technologies that actively and intentionally engender deliberation and provoke it in cases in which public debate has reached a stalemate. With respect to genomics, deliberatively engaged people could ask themselves this question: which technology/innovation engenders the most interesting and innovative deliberation?

Conclusion

Deliberative participation can imply mapping different arrangements of the interaction between genomics and societal developments. I began this chapter with an overview of the most important theoretical issues regarding the deliberative approach. I then discussed five challenges: problems of unequal participation, cognitive and normative uncertainties, different scripts, multi-level problems, and finally the deliberative role of deliberation-eliciting technologies. Discussing these challenges, I developed tools from a non-traditional ethical perspective that included imaginary futures through aesthetic explorations, future explorations using scenarios of different moral screenplays and dramatic rehearsals, accountable leadership, and bridge organizations. With these tools, the deliberative ethicist develops as a moral entrepreneur, not only as a communicator who facilitates deliberative exchange.

NOTE

This study is based on my experiences with deliberations in many research projects on the ethical aspects of genomics, mostly financed by the Dutch Science Organization. I wish to thank collaborators for their inspiring ideas and work, in particular, Jozef Keulartz, Henk van den Belt, Bram de Jonge, and Clemens Driessen.

REFERENCES

Akrich, M. 1992. "The De-scription of Technical Objects." In *Shaping Technology/ Building Society*, edited by W.E. Bijker and J. Law, 205-24. Cambridge, MA: MIT Press.

Benhabib, S., ed. 1996. *Democracy and Difference: Contesting the Boundaries of the Political.* Princeton: Princeton University Press.

Bohman, J. 2007. *Democracy across Borders: From Demos to Demoi.* Cambridge, MA: MIT Press.

Bryson, B. 2003. *A Short History of Nearly Everything.* London: Black Swan.

Burgess, M.M. 2003. "What Difference Does Public Consultation Make to Ethics?" Electronic Working Papers Series, W. Maurice Young Centre for Applied Ethics, University of British Columbia. http://www.ethics.ubc.ca.

Chambers, S. 2003. "Deliberative Democratic Theory." *Annual Review of Political Science* 6: 307-26.

Delli Carpini, M., F. Cook, and L. Jacobs. 2004. "Public Deliberation, Discursive Participation, and Citizen Engagement: A Review of the Empirical Literature." *Annual Review of Political Science* 7: 315-44.

Driessen, C., and M. Korthals. Forthcoming. "Technology Development as Disclosing Moral Worlds: Creating Ethical Learning Processes by Design." *Social Studies of Science.*

Dryzek, J.S. 1990. *Discursive Democracy: Politics, Policy, and Political Science.* New York: Cambridge University Press.

–. 2000. *Deliberative Democracy and Beyond.* Oxford: Oxford University Press.

Einsiedel, E., and D. Eastlick. 2000. "Consensus Conferences as Deliberative Democracy: A Communications Perspective." *Science Communication* 21, 4: 323-43.

Feenberg, A. 2001. "Democratizing Technology: Interests, Codes, Rights." *Journal of Ethics* 5: 177-95.

Gastil, J., and P. Levine, eds. 2005. *The Deliberative Democracy Handbook.* New York: Wiley.

Goffman, E. 1974. *Frame Analysis: An Essay on the Organization of Experience.* Cambridge, UK: Cambridge University Press.

Goodin, R., and J. Dryzek. 2006. "Deliberative Impacts: The Macro-Political Uptake of Mini-Politics." *Politics and Society* 34, 2: 219-44.

Habermas, J. 1971. "Vorbereitende Bemerkungen zu einer Theorie der kommunikativen Kompetenz." In *Theorie der Gesellschaft oder Sozialtechnologie,* by J. Habermas, 101-41. Frankfurt: Suhrkamp.

–. 1996. *Between Facts and Norms: Contributions to a Discourse Theory of Law and Democracy.* Cambridge, MA: MIT Press.

–. 1998. "Habermas Responds to His Critics." In *Habermas on Law and Democracy,* edited by M. Rosenfeld and A. Arato, 381-453. Berkeley: University of California Press.

Hajer, M., and H. Wagenaar, eds. 2003. *Deliberative Policy Analysis: Understanding Governance in the Network Society.* Cambridge, UK: Cambridge University Press.

Hamlett, P. 2003. "Technology Theory and Deliberative Democracy." *Science, Technology, and Human Values* 28, 1: 112-40.

Keulartz, J., M. Korthals, M. Schermer, and T. Swierstra. 2002. *Pragmatist Ethics for a Technological Culture.* Dordrecht: Kluwer

Keulartz, J., and G. Leistra, eds. 2008. *Legitimacy in European Nature Conservation Policy: Case Studies in Multilevel Governance.* International Library of Environmental, Agricultural, and Food Ethics 14. Springer: Dordrecht.

Keulartz, J., M. Schermer, M. Korthals, and T. Swierstra. 2004. "Ethics in a Technological Culture: A Programmatic Proposal for a Pragmatist Approach." *Science, Technology, and Human Values* 29, 1: 3-30.

Kitcher, P. 2002. *Science, Truth, and Democracy.* Oxford: Oxford University Press.

Komduur, R., M. Korthals, and H. te Molder. 2008. "The Good Life: Living for Health and a Life without Risks? On a Prominent Script of Nutrigenomics." *British Journal of Nutrition* 101, 3: 307-17.

Korthals, M. 2008a. "Ethics and Politics of Food: Toward a Deliberative Perspective." *Journal of Social Philosophy* 39, 3: 445-63.

–. 2008b. "Ethical Rooms for Maneuver and Their Prospects vis-à-vis the Current Ethical Food Policies in Europe." *Journal of Agricultural and Environmental Ethics* 21: 249-73.

Korthals, M., and R. Komduur. 2010. "Uncertainties of Nutrigenomics and Their Ethical Meaning." *Journal of Agricultural and Environmental Ethics* 23: 211-26.

Kuhn, T., 1962, *The Structure of Scientific Revolutions*, Chicago: Chicago University Press.

Merton, R.K. 1968. "Science and the Social Order." In *Social Theory and Social Structure*, by R.K. Merton, 591-603. New York: The Free Press.

Mouffe, C., ed. 1996. *Dimensions of Radical Democracy: Pluralism, Citizenship, Community.* London: Verso.

Pearson, H. 2006. "Genetics: What Is a Gene." *Nature* 441: 398-401.

Peterson, G.D., G.S. Cumming, and S.R. Carpenter. 2003. "Scenario Planning: A Tool for Conservation in an Uncertain World." *Conservation Biology* 17: 358-66.

Piatigorsky, J. 2007. *Gene Sharing and Evolution.* Cambridge, MA: Harvard University Press.

Renn, O., and T. Webler. 1996. "Der kooperative Diskurs: Grundkonzeptionen und Fallbeispiel." *Analyse und Kritik* 18: 175-207.

Rowe, G., and L.J. Frewer. 2004. "Evaluating Public Participation Exercises: A Research Agenda." *Science, Technology, and Human Values* 29, 4: 512-56.

Rowe, G., R. Marsh, and L.J. Frewer. 2004. "Evaluation of a Deliberative Conference Using Validated Criteria." *Science, Technology, and Human Values* 29, 1: 88-121.

Scharpf, F.W. 1999. *Governing in Europe: Effective and Democratic?* Oxford: Oxford University Press.

Schot, J. 2001. "Towards New Forms of Participatory Technology Development." *Technology Analysis and Strategic Management* 13, 1: 39-52.

Sriram, G., J. Martinez, E. McCabe, J. Liao, and K. Dipple. 2005. "Single-Gene Disorders: What Role Could Moonlighting Enzymes Play?" *American Journal of Human Genetics* 76, 6: 911-24.

Theiss-Morse, E. 2002. "The Perils of Voice and the Desire for Stealth Democracy." *Maine Policy Review* 11: 80-89.

Theiss-Morse, E., and J.R. Hibbing. 2005. "Citizenship and Civic Engagement." *Annual Review of Political Science* 8: 227-49.

Understanding Risk Team. 2004. *A Deliberative Future? An Independent Evaluation of the GM Nation.* Cardiff: Cardiff University.

PART 3

INTERNAL CONDITIONS FOR LEGITIMATE PUBLIC ENGAGEMENT
Lessons for the Practitioner

Deliberative Fears
Citizen Deliberation about Science in a National Consensus Conference

MICHAEL D. COBB

In this chapter, I take a critical look at how the science and technology (S&T) literature has portrayed the practice and value of structured citizen deliberation. Although S&T scholars increasingly promote citizen deliberation as a desirable method of engaging citizens with scientific issues and generating informed public opinion about them, it is not obvious that these efforts have been fruitful. There are numerous reasons for my uncertainty, and here I identify three of them: the literature frequently blurs the crucial conceptual difference between having citizens *talk about science* and having them *deliberate about it;* it typically ignores criticisms of deliberation that are prominent in other disciplines; and it only occasionally presents empirical evidence measuring the presence and outcomes of deliberation.

To be clear, raising concerns about the state of the literature does not mean I reject the notion that deliberation might benefit citizens and encourage them to participate in their own governance. Indeed, I have found that after deliberating, citizens develop informed policy preferences and become more efficacious and engaged with issues of science and technology (Cobb 2011). They also endorse the process when asked about it more than a year afterwards (Cobb and Gano, forthcoming). Rather, I see this as an opportunity to reiterate the need for more frequent and rigorous empirical assessments of deliberation to more reliably evaluate the merits of its adoption (Burgess and Chilvers 2006; Rowe, Marsh, and Frewer 2004). But to do that, future research also needs to demonstrate empirically that the phenomenon being studied is actually deliberation and not some other form of communication. A serious problem in the literature is that it is based on referencing outcomes of interpersonal communication that have questionable links to definitions of deliberation (on this point, see Sprain and Gastil 2006). So, in an admittedly modest contribution to the literature, I also present results from a study of a national consensus at which I indirectly measured deliberation. Although the measures turned out to be problematic, it is important to continue trying to measure deliberation to better understand its value as a method of citizen engagement with science.

Background

Scholars have painted a rather dismal picture of Americans' knowledge of worldly affairs in general and issues such as science and technology in particular (Delli Carpini and Keeter 1996; National Science Board 2010; Nisbet and Scheufele 2009). Yet, many scholars advocate for the democratization of science so that ordinary citizens can influence the development trajectories of emerging technologies (Macnaghten, Kearnes, and Wynne 2005; Powell and Colin 2009; Wilsdon, Wynne, and Stilgoe 2005). Obviously, these conflicting perspectives create a puzzle: how can ordinary citizens participate in complex decision-making processes when their competence is routinely called into question?

A growing number of scholars have proposed as a potential solution establishing deliberative processes, such as deliberative polling, in which citizens can become engaged with and informed about an issue (Fishkin 1988; Fishkin and Luskin 1996). In the S&T literature, often the specific format is a consensus conference (Cobb 2011; Einsiedel and Eastlick 2000; Guston 1999; Hamlett and Cobb 2006; Sclove 1995). These processes are meant to bridge the gap between uninformed public opinion and the representative bodies that supposedly, but perhaps fail to, account for citizens' preferences. The preference for promoting citizen deliberation compared with other forms of communication is rooted in democratic theory. In short, its advocates claim that deliberation results in better citizens, better judgments, and more legitimate outcomes (Ackerman and Fishkin 2004; Barber 1984; Dryzek 2000; Gutman and Thompson 2004; Gastil and Levine 2005; Habermas 1996; Luskin, Fishkin, and Jowell 2002; Thompson 2008). I review these claims, and criticisms of deliberation, in the next section.

Deliberative Hopes and Fears

In the S&T literature, deliberation as a method for engaging citizens with science is widely applauded, and democratic theory seems to legitimize it. Yet, the literature on democratic theory and deliberation is not monolithic, and criticisms of deliberation are actually prevalent (e.g., Mutz 2008). Only recently have critical perspectives on deliberation appeared in the S&T literature, and even these studies tend to find fault with particular aspects of a specific deliberative event rather than questioning the notion that citizens should deliberate more (Delborne et al. 2011; Kleinman, Delborne, and Anderson 2011; Philbrick and Barandiaran 2009; Powell and Colin 2009). As a result, the S&T literature has arguably developed an overly optimistic perspective on the potential for deliberation to improve both citizens and policies.

Deliberative Hopes

Advocates of greater citizen deliberation emphasize two main hypothesized virtues. The first one is that deliberation produces better decisions. Primarily, this means that collective decisions are better than they would have been absent prior deliberation. Rawls (1971, 358-59), for example, writes that "discussion is a way of combining information and enlarging the range of arguments. Each person can share what he or she knows with the others, making the whole at least equal to the sum of the parts." To be sure, it is difficult to measure better group decisions because doing so requires an independent, objective standard against which to evaluate the deliberative decision. Beyond things that we know with certainty, which could provide a benchmark for "correct" or "incorrect" decisions, we normally lack appropriate measures. Instead, outcomes are labelled "better" because they incorporate the normatively based belief that the process of inclusion in which everyone is equally capable of affecting decisions confers greater legitimacy on binding decisions (Gutman and Thompson 2004).

Better decisions, however, are measured at the individual level. As Fishkin argues (1991, 1997; see also Ackerman and Fishkin 2004; Luskin, Fishkin, and Jowell 2002), deliberation foremost produces informed opinions. Pidgeon and colleagues (2009) describe the benefit of deliberating on nanotechnology in terms of the coherent opinions generated that surveys cannot measure. Likewise, Esterling, Neblo, and Lazer (2011) find that citizens become more knowledgeable about politically controversial issues as a result of interacting with their political representatives and then deliberating with their peers about that communication. Individual preferences are expected to become more congruent with objectively defined interests when one is more informed. In this account, surveys that measure shifting opinions after deliberation identify the process of individuals coming to better understand how their own interests are affected by different courses of policy action (Barabas 2004).

Another presumed benefit of deliberation is that it produces better citizens. Deliberation is thought to encourage political participation, broadly understood, and increase social capital in particular (Putnam 2000). The components of social capital, such as knowledge, efficacy, trust, and civility, are valued because they are theoretical prerequisites of a healthy democracy. Although easier to measure, data on these kinds of benefits is still rare (Delli Carpini, Cook, and Jacobs 2004). Morrell (2005), though, finds some evidence that participation in deliberation can have positive effects on political efficacy. In a similar vein, Gastil and colleagues (2010) find that participation in jury duty, a classic deliberative environment, leads to increased voter participation. My recent study of deliberation in consensus conferences finds evidence of increased knowledge, trust, and internal efficacy but also reduced external efficacy (Cobb 2011).

Looking across the different methodologies of these studies, however, one notices that many of them are not really comparable, and it is questionable that they are describing the effects of deliberation. Consider deliberative polling, for instance, which occurs over a single weekend (Fishkin and Luskin 1996). Critics have argued since its inception that it is in reality neither opinion polling nor deliberation (see Merkle 1996). Or consider a study that recruits citizens to pose questions to their political representative online and then has them discuss the issue in a web-based chat room for twenty-five minutes (Esterling, Neblo, and Lazer 2011). Are twenty-five minutes of online conversation sufficient to generate deliberation? What about having college students discuss issues in a single class (Morrell 2005)? Does deliberation occur during a half-day workshop (Pidgeon et al. 2009)? Although it is not obvious precisely how much time is necessary for deliberation to take place, one needs to recognize that these so-called deliberative events do not offer any direct evidence that deliberation actually occurred.

Deliberative Fears
Despite the hopes for deliberation, critics have assembled an impressive list of its supposed flaws (Hibbing and Theiss-Morse 2002; Sanders 1997;

Stokes 1998; Sunstein 2000, 2003, 2005; see Mendelberg 2002 for a thorough review). One perspective is what Luskin, Fishkin, and Hahn (2007) call *defeatist,* which holds that deliberative democracy is unattainable and unnecessary. As some argue, citizens already arrive at reasoned judgments through an efficient use of cognitive shortcuts (Lupia 1994; Posner 2005). A second perspective is that undesirable outcomes are a routine consequence of citizen deliberation. For these scholars, the worry is that privileged individuals will dominate proceedings to their own advantage. Emphasizing this point, Hibbing and Theiss-Morse (2002, 184) write that "increased interaction will not boost political capital at all and may very well do damage." A third and the most common critique is that decision-making biases are exaggerated by deliberation. As I elaborate in the next section, deliberation is faulted for two (ironically) contradictory pathologies found in public opinion research: the tendency of individuals to resist new yet pertinent information and remain wedded to their original beliefs, and their tendency to readily adopt the positions of the majority, regardless of the veracity of those positions. In the first case, deliberation is a waste of time. In the second, it is worse than not deliberating at all because people adopt the positions held by the majority due to social pressures and the original numerical distribution of preferences rather than the quality and merits of the majority's arguments.

Fear of Attitudinal Consistency

As a useful analogy, self-interested people are assumed to have firm opinions, and arenas for debates are thought to be merely for show or, worse, a premeditated ruse to give the appearance that persuasion is possible when it is not. Consistency of opinion is also often motivated by concerns about reputation (Sunstein 2000, 2003). Whatever the reasons, rational choice theory posits consistency of beliefs among actors, even in the face of new and possibly disconfirming information. Deliberation in these kinds of settings is not intended to change minds, nor is it expected to.

Research on the dynamics of public opinion more generally reaches a similar conclusion. Drawing on the theory of motivated reasoning (Kunda 1990), scholars fail to find belief change in cases in which logically it should occur (Bullock 2007; Cobb 2007). Individuals, for example, stubbornly cling to their original positions even when confronted with factual data to the contrary (Nyhan and Reifler 2010). Studies also demonstrate citizens' selective perception of and attention to pleasing or threatening information, meaning that citizens distort or fail to hear opposing sides of a debate (Iyengar and

Hahn 2009). In short, abundant research finds that people respond to new information by counterarguing against contrary ideas instead of doing what democratic theory requires – reconciling new information with old.

Fear of Polarization

It seems that, in reaction to the above evidence, scholars prefer people to be open to persuasion. This is not necessarily the case. The downsides to openness are said to be numerous. As recent studies demonstrate, individuals are willing to be influenced by irrational or even blatantly false information (Kosloff et al. 2010). Alternatively, more talk about controversial issues can lead to risk amplification (Binder et al. 2010). Perceptions of risk can increase through deliberation because people tend to converse mainly within circles of like-minded individuals. Their conversations are characterized by a skewed pool of information that reinforces the discussants' original beliefs.

Likewise, scholars are concerned about the widespread susceptibility of individuals within small groups to polarization effects (Sunstein 2003).[1] Polarization effects occur when individuals holding the minority opinion in a group adopt the majority opinion after deliberating, not because of the merits of the majority's ideas but because of the numerical advantage of those ideas within the group. As Sunstein warns, groups consisting of individuals with extreme tendencies are likely to shift, and they are likely to shift more without repeated exposure to competing views. Worse, rather than fostering an exchange of information that is weighed carefully, deliberation serves merely to benefit the already advantaged by amplifying their opinions due to greater cognitive and experiential resources (Mendelberg 2002; Sanders 1997).[2] If group deliberation merely serves to ratify the preferences of the well off, then it is perhaps worse than not deliberating at all.

A second reading of these complaints, however, reveals the same concern that I raised about research that indicates the benefits of deliberation. Namely, I am uncertain whether the phenomenon being critiqued is deliberation. Sunstein's (2005) criticisms, for example, are typically based on his interpretations of social psychology experiments on group dynamics that were not designed to study deliberation properly defined. Polarization indeed seems to occur more readily among "like-minded people," but democratic deliberation invokes the need for heterogeneous participant pools. Thus, it appears that many criticisms of deliberation in the literature depend on references to group *discussion,* but that is not the same thing as *deliberation.* Without demonstrating that deliberation occurred, it is highly questionable to presume that talking during any kind of citizen engagement automatically qualifies as deliberation. Many activities requiring people to talk about controversial issues are obviously polarizing, but that does not mean that deliberation is polarizing.

What Is Deliberation?

Defining deliberation is essential to analyzing it, yet there is no singular established definition that permeates the literature. Instead, there are many definitions of it, some having more stringent rules for what constitutes deliberative talk from mere argumentation. This makes moving from theory to empiricism problematic, to say the least (Mutz 2008). In the more egregious instances, however, mere talking has been sufficient to be considered deliberation. As a result, I argue that evaluations of deliberation in the S&T literature are often not about deliberation at all.

Clarifying the Concept

Both the act of deliberating and its consequences are, in theory, measurable. How can we know whether talk amounts to deliberation? At a simple level, it is universally accepted that deliberation involves a particularly sophisticated version of talking, listening, and reasoning. Deliberation is also said to be a multi-layered process or at least a process defined by multiple criteria. For example, Burkhalter, Gastil, and Kelshaw (2002) describe deliberation alternatively as a form of governance and a type of communication. Here I am only interested in unpacking deliberation as a form of communication.

Mendelberg (2002) provides a welcome starting point for measuring deliberation as it is actually practised. Foremost, she writes, deliberation is about talking, and it is inherently about groups.[3] Clearly, talking takes place in small groups because deliberation among the entire polity simultaneously is inconceivable. Another reason for making small groups the focus of analysis is that scholars are pessimistic about being able to successfully engage the public as a whole: many people simply do not care one way or another about these kinds of issues (Nisbet and Scheufele 2009). I also agree with Delli Carpini, Cook, and Jacobs (2004) when they say that "ordinary citizens" do the talking. Talking becomes deliberative, says Habermas (1996), when argumentation is welcomed and it is based on a free and equal exchange. More than that, deliberative talk requires individuals to weigh carefully both the consequences of various options for action and how others will view these acts (Burkhalter, Gastil, and Kelshaw 2002). Finally, as Ryfe (2002) explains, deliberation occurs when claims are advanced, evidence is presented, and counterfactuals are considered.[4]

Integrating these various themes, though certainly not definitive, I find that deliberation is, minimally, a process whereby ordinary citizens of diverse backgrounds (i.e., opinions) freely and cognitively engage in an exchange of ideas, beliefs, and knowledge about policy. Yet these are admittedly necessary but insufficient conditions for deliberation. Importantly, an implicit level of cognitive activity is required when explaining one's preferences and when considering others' viewpoints. Mere acknowledgement that one has heard the other person is too passive to be deliberative. Deliberation requires offering explanations and evaluating competing claims. Cognitive engagement leads to integrating claims and counterclaims into new formulations of positions and arguments (Macoubrie 2003).

I further add the label "democratic" to deliberation defined above when certain rules are present to guide deliberation. In deliberative events, access to information needs to be unencumbered, the overall process needs to be transparent to participants and observers, and subsequent group decisions must be capable of affecting policy. In general, the rules are congruent with the five criteria of democratic rules laid out by Rowe and Frewer (2000): representativeness of participants, early involvement in an issue cycle, potential for policy impact, process transparency, and resource accessibility.

Talking versus Deliberating

To reiterate, I am skeptical about how much we know about the effects of citizen deliberation on science and technology (but see Cobb 2011; Hamlett and Cobb 2006; Powell and Kleinman 2008). The mistake of conflating talking with deliberation is not unique to the S&T literature, but it appears to be more common. To illustrate, Burgess and Chilvers (2006) provide a typology of engagement strategies, ranging from education to deliberation. As communication in the methods to promote the activity moves from a one-way transmission of information to two-way interactions, it eventually becomes deliberative. The methods appropriate to these varying levels of communication are not the same. In the case of education, for example, information is passively distributed by leaflets, exhibitions, and media. In the case of citizen deliberation, however, models such as citizen juries and consensus conferences that demand cognitive engagement are necessary to generate and sustain it. These resource-intensive methods are required because their structure is geared toward meeting the required conditions for deliberative talk, though of course they do not guarantee it. Thus, the nature of talking in deliberative events needs to be measured.

Survey Measures of Deliberation in a National Consensus Conference

Moving from theory to empiricism, I present results on deliberation in a national consensus forum in the United States about human enhancement technologies (the NCTF).[5] Since consensus conferences have been described in great detail elsewhere (see Einsiedel and Eastlick 2000; Einsiedel, Jelsøe, and Breck 2001; Grundhal 1995; Guston 1999; Powell and Kleinman 2008), and several in-depth explanations of the NCTF process have been written (see Hamlett, Cobb, and Guston 2008; Wickson, Cobb, and Hamlett 2012), I avoid unnecessarily repeating their fuller descriptions here.

In brief, the NCTF was held simultaneously in six distinct US cities in March 2008. Over 350 citizens applied, and fifteen per site were ultimately selected to participate. To facilitate deliberation across space, we utilized the Internet as a mode of deliberative interaction to overcome geography (Delborne et al. 2011).[6] Procedurally, participants deliberated face to face in their respective geographic groups for one weekend at the beginning of the month, and they deliberated electronically across their geographic groups in nine two-hour sessions during the rest of the month. To promote deliberation, citizens took part in question-and-answer sessions with a diverse group of topical experts, and facilitators made sure that all participants had opportunities to speak and ask questions.

Measures

To evaluate deliberation in the NCTF, participants were required to complete a lengthy pre-test and post-test questionnaire. The pre-test was taken online before their first face-to-face meeting, and the post-test was taken online soon after the groups generated their consensus reports. Although all but one participant completed the pre-test, some data were lost because four people dropped out before the first meeting and twelve panelists did not complete the post-test. Besides trying to measure deliberation, the surveys measured participants' knowledge of and opinions about nanotechnology and human enhancement and their self-reported feelings of efficacy and trust in others (Cobb 2011; Hamlett, Cobb, and Guston 2008).

Deliberation

Normally, deliberation is presumed to occur based on the specific processes and structures in place when citizens talk about an issue (Hamlett and Cobb 2006). More recently, scholars have tried content analysis approaches to actually measure the quality of deliberation (Black et al. 2009; Steenbergen et

al. 2003). These methods, however, are either labour intensive or better suited to face-to-face deliberations, where conversations are synchronous. With
online deliberation, however, comments and reactions to them are often
asynchronous. Instead, following Macoubrie (2003), I constructed survey
questions to measure a person's *willingness* to do the kinds of things demanded by definitions of deliberation. Participants' willingness to deliberate was measured by self-reported agreement with statements that define
deliberative activities. Of course, this is an imperfect measurement strategy
for a variety of reasons. Not only am I indirectly measuring deliberation, but
also, I am probably overestimating the willingness to deliberate due to social
desirability pressures. Despite these flaws, this method permits at least a
macro-level analysis of attitudes before and after the NCTF.

More specifically, participants were asked how accurately six items described their feelings about communication with others (see Appendix for
question wording). Each answer was recorded on a six-point scale, ranging
from 1 (very inaccurate) to 6 (very accurate). As indicated, these statements
tried to capture acceptance of procedural rules for deliberation and the difference between proto-deliberative tendencies (i.e., exchanging views) and
fully deliberative ones (i.e., integrating information to develop solutions)
(see Macoubrie 2003). For example, a proto-deliberative measure states
that, "when I disagree with someone, trying to understand that person's reasoning is important to me." Conversely, an example of a fully deliberative
statement reads: "I am willing to adjust my positions on issues if I hear good
reasons to do so."

Results

Unfortunately, subsequent factor analysis failed to distinguish between the
presumed lower and higher levels of deliberation. Panelists' answers were
instead arrayed on two dimensions representing the principles required of
deliberation and the endorsement of types of discussion rules (majority rule
versus consensus). As a result, I created two additive indices, one for the
two items about rules and another for the four items measuring deliberative
principles. To reiterate, these items were presented before and after the
panelists took part in the NCTF. The pre-test and post-test Cronbach's
alphas for the two-item indices were, respectively, .53 and .63. Alphas for
the four-item scales in the pre-test and post-test were, respectively, .58 and
.74. Overall, these are modestly acceptable scale values.

I present results for the sample as a whole because behaviour at each site
location was nearly identical.[7] Before participating in the NCTF, most

panelists reported high levels of deliberative tendencies. The mean score for the two-item scale measuring support for consensus rules was 8.5 when the maximum possible score was 12 (the median score was 9). Likewise, the average score for the four-item index was 20.7 when the maximum score was 24 (the median score was 21). Although these measures failed to distinguish between panelists' willingness to deliberate, they do support my belief that deliberation took place in the NCTF. Not only were the processes linked to promoting deliberation present, but also, nearly every participant endorsed the conference rules and types of communicative activities that result in deliberation.

One other problem, however, is that the over-time analysis of participants' willingness to deliberate is hampered by ceiling effects. Since initial scores were so high, it is nearly impossible to learn whether participants become *more willing* to deliberate after actually taking part in a consensus conference. Indeed, the scores for both indices remained virtually identical at the end of the conference. The average post-test scores on the two-item and four-item indices at the end of it were, respectively, 8.6 and 20.6. Considering that the willingness to deliberate was nearly universal at the onset of the conference, it is worth pointing out that at least average scores *did not decline,* as would be predicted by Hibbing and Theiss-Morse (2002), who contend that people do not like to deliberate.

It is possible, though, that analysis of the two deliberation indices obscures important individual survey item differences. I examined this possibility by disaggregating the indices and comparing answers for each of the six scale items. This analysis revealed two interesting results. First, I found that agreement with consensus rules increased over time ($p<.05$), but opinions about majority rules did not (*ns*). Second, there was less endorsement of one specific principle of deliberation: "I am willing to adjust my positions on issues if I hear good reasons to do so ($p<.05$)." I probed this finding a little further and discovered that it was driven almost entirely by opinions at a single site location (Colorado). Although the negative opinion change at one site offers some support for critics' accusation that deliberation begets discontent, this one result was clearly unrepresentative of NCTF outcomes.

Discussion

In this chapter, I levelled several criticisms at the literature discussing citizen deliberation about science and technology. From my perspective, the literature lumps too many disparate engagement events together when describing deliberation, and it then fails to adequately evaluate the outcomes of deliberative events, in large part because it ignores criticisms of deliberation in other disciplines. It is important to get a better understanding of the outcomes of citizen deliberation because significant resources are being devoted to engaging citizens with science. But to do that, we first need to do a better job of defining the concept to distinguish it from other forms of group discussion.

I also described a simple process for measuring deliberation by investigating how citizens think about principles that define deliberation. Although I do not directly measure deliberation, the survey measures that I crafted can be useful. For instance, researchers can examine the scores at the onset of an engagement activity and assess the probability that citizens are prone to deliberating. Or, by comparing the score values before and after an engagement event, researchers can evaluate whether citizens increasingly value the criteria of deliberation. Greater endorsement of deliberative principles after an engagement event that utilizes processes theorized to promote deliberation, for example, is additional evidence to support the claim that deliberation occurred.

Going forward, by defining and measuring the concept of deliberation, it will be possible to evaluate more accurately whether having citizens deliberate, not just briefly discuss policy, is a necessity or merely a luxury. Are consensus conferences superior to other kinds of less resource-intensive engagement activities? Other than referencing theory, at the moment we cannot really answer such a question. Likewise, future studies of deliberation in consensus conferences should also build comparison groups into their research designs. In the best cases, deliberation might lead to normatively desirable outcomes, but it is not obvious that structured deliberation should be preferred to other types of engagement or even to doing nothing at all. For example, methodologies for testing new drugs include a control group that receives no treatment and a second treatment group that receives a placebo. The validity of new drugs depends not just on outperforming the absence of treatment but also on beating the performance of a product unintended to be effective. Do deliberative events outperform less resource-intensive engagement exercises, such as science cafés? Future research is clearly needed to wrestle with these issues.

Appendix: Question Wording for Deliberation Scale

Please read the following statements and indicate how well they describe you. Using a scale from 0 to 5, where "0" means the statement you read is very inaccurate and "5" means the statement you read is very accurate, choose the appropriate number.

Q1. When I disagree with someone, trying to understand that person's reasoning is important to me.

Q2. When I disagree with someone, helping that person understand my reasoning is important to me.

Q3. I prefer to let the majority decide what to do when group conflicts arise (reverse coded).

Q4. I prefer to reach a group consensus when group conflicts arise.

Q5. I am motivated to try to find common ground with people who disagree with me.

Q6. I am willing to adjust my positions on issues if I hear good reasons to do so.

NOTES

I am grateful for feedback provided by Patrick Hamlett and David Guston, NCTF collaborators at six site locations, and participants at the workshop Publics and Emerging Technologies: Cultures, Contexts, and Challenges, Banff, AB, 30-31 October 2009. Preparation of this chapter was supported by the Center for Nanotechnology in Society at Arizona State University (NSF Grant 0531194). The views expressed here are mine alone and do not necessarily represent those of the National Science Foundation.

1 Asch's (1956) path-breaking experiments on conformity have led to decades of further confirmatory studies warning that people are all too willing to accept obviously erroneous judgments, presumably motivated by the simple desire to "belong" to the group.

2 Sunstein (2005) notes that, though polarization effects are possibly natural and quite robust, institutions might be designed to prevent them. Nevertheless, he also argues that the absence of institutions in real-world deliberations is the norm rather than the exception, and thus, routine public group deliberations are likely to degenerate into extreme points of view.

3 Goodin (2000) argues that deliberation should also be thought of as something that takes place within individuals. Deliberation has an "internal reflective" aspect to it, just as much as it obviously has an "external-collective" nature. Shapiro (2002), however, argues that deliberation is not an isolated activity. At the individual level, one can only be reflective, not deliberative.

4 Another possible element of deliberation that I do not take up here is the role of emotion. Classical notions of deliberation highlight reason (Habermas 1979), while recent conceptions suggest that emotions are integral to the process (Gastil 2000; Mendelberg 2002).

5 Examples of enhancements include applications such as implants to allow direct computer-to-brain links or medical devices that roam the blood stream searching for cancer cells (Hays, Cobb, and Miller 2013).
6 Of course, it is not possible, strictly speaking, to achieve representativeness of all community demographics within small groups, but the NCTF panelists overall reflected a reasonable approximation of the American public on important demographic characteristics (Cobb 2011; Hamlett, Cobb, and Guston 2008).
7 Participants at the Colorado site reported significantly lower endorsement of deliberative principles after deliberating, a finding supported by anecdotal evidence provided by site organizers.

REFERENCES

Ackerman, B., and J.S. Fishkin. 2004. *Deliberation Day*. New Haven: Yale University Press.

Asch, S.E. 1956. "Studies of Independence and Conformity: A Minority of One against a Unanimous Majority." *Psychological Monographs* 70: 1-70.

Barabas, J. 2004. "How Deliberation Affects Policy Opinions." *American Political Science Review* 98: 687-701.

Barber, B. 1984. *Strong Democracy*. Berkeley: University of California Press.

Binder, A.R., D.A. Sheufele, D. Brossard, and A.C. Gunther. 2010. "Interpersonal Amplification of Risk? Citizen Discussions and Their Impact on Perceptions of Risks and Benefits of a Biological Research Facility." *Risk Analysis* 31, 2: 324-34.

Black, L.W., S. Burkhalter, J. Gastil, and J. Stromer-Galley. 2009. "Methods for Analyzing and Measuring Group Deliberation." In *Sourcebook of Political Communication Research: Methods, Measures, and Analytical Techniques*, edited by L. Holbert and E. Bucy, 323-45. New York: Routledge.

Bullock, J. 2007. "Experiments on Partisanship and Public Opinion: Party Cues, False Beliefs, and Bayesian Updating." PhD diss., Stanford University.

Burgess, J., and J. Chilvers. 2006. "Upping the *Ante:* A Conceptual Framework for Designing and Evaluating Participatory Technology Assessments." *Science and Public Policy* 33: 713-28.

Burkhalter, S., J. Gastil, and T. Kelshaw. 2002. "A Conceptual Definition and Theoretical Model of Public Deliberation in Small Face-to-Face Groups." *Communication Theory* 12: 398-422.

Cobb, M.D. 2007. "Knowing the Truth Is Not Enough: The Resilience of Discredited Information." Paper prepared for presentation at the 2007 annual meeting of the International Society of Political Psychology, Portland, OR, 4-7 July.

–. 2011. "Creating Informed Public Opinion: Citizen Deliberation about Nanotechnologies for Human Enhancements." *Journal of Nanoparticle Research* 13: 1533-48.

Cobb, M.D., and G. Gano. Forthcoming. "Evaluating Structured Deliberations about Emerging Technologies: Post-Process Participant Evaluation." *International Journal of Emerging Technologies*.

Delborne, J.A., A.A. Anderson, D.L. Kleinman, M. Colin, and M. Powell. 2011. "Virtual Deliberation? Prospects and Challenges for Integrating the Internet in Consensus Conferences." *Public Understanding of Science* 20: 367-84.

Delli Carpini, M.X., F.L. Cook, and L. Jacobs. 2004. "Public Deliberation, Discursive Participation, and Citizen Engagement: A Review of the Empirical Literature." *Annual Review of Political Science* 7: 315-44.

Delli Carpini, M.X., and S. Keeter. 1996. *What Americans Know about Politics and Why It Matters.* New Haven: Yale University Press.

Dryzek, J.S. 2000. *Deliberative Democracy and Beyond: Liberals, Critics, Contestations.* Oxford: Oxford University Press.

Einsiedel, E.F., and D.L. Eastlick. 2000. "Consensus Conferences as Deliberative Democracy." *Science Communication* 21: 323-43.

Einsiedel, E.F., E. Jelsøe, and T. Breck. 2001. "Publics at the Technology Table: The Consensus Conference in Denmark, Canada, and Australia." *Public Understanding of Science* 10: 83-98.

Esterling, K.M., M.A. Neblo, and D.M. Lazer. 2011. "Means, Motive, and Opportunity in Becoming Informed about Politics: A Deliberative Field Experiment with Members of Congress and Their Constituents." *Public Opinion Quarterly* 75: 483-503.

Fishkin, J. 1988. "The Case for a National Caucus: Taking Democracy Seriously." *Atlantic Monthly,* August, 16-18.

–. 1991. *Democracy and Deliberation: New Directions for Democratic Reform.* New Haven: Yale University Press.

–. 1997. *The Voice of the People: Public Opinion and Democracy.* New Haven: Yale University Press.

Fishkin, J., and R. Luskin. 1996. "The Deliberative Poll: A Reply to Our Critics." *Public Perspective* 7: 45-49.

Gastil, J. 2000. *By Popular Demand: Revitalizing Representative Democracy through Deliberative Elections.* Berkeley: University of California Press.

Gastil, J., P. Deess, P.J. Weiser, and C. Simmons. 2010. *The Jury and Democracy: How Jury Deliberation Promotes Civic Engagement and Political Participation.* Oxford: Oxford University Press.

Gastil, J., and P. Levine, eds. 2005. *The Deliberative Democracy Handbook: Strategies for Effective Civic Engagement in the 21st Century.* San Francisco: Jossey-Bass.

Goodin, R. 2000. "Democratic Deliberation Within." *Philosophy and Public Affairs* 29: 81-109.

Grundhal, J. 1995. "The Danish Consensus Conference Model." In *Public Participation in Science: The Role of Consensus Conferences in Europe,* edited by S. Joss and J. Durant, 31-41. London: Science Museum.

Guston, D. 1999. "Evaluating the First U.S. Consensus Conference: The Impact of the Citizens' Panel on Telecommunications and the Future of Democracy." *Science, Technology, and Human Values* 24: 451-82.

Gutman, A., and D. Thompson. 2004. *Why Deliberative Democracy?* Princeton: Princeton University Press.

Habermas, J. 1979. Communication and the Evolution of Society, T. McCarthy, trans. London: Heinemann.

—. 1996. *Between Facts and Norms: Contributions to a Discourse Theory of Law and Democracy.* Translated by W. Rehg. Cambridge, MA: MIT Press.

Hamlett, P., and M.D. Cobb. 2006. "Potential Solutions to Public Deliberation Problems: Structured Deliberations and Polarization Cascades." *Policy Studies Journal* 34: 629-48.

Hamlett, P., M.D. Cobb, and D. Guston. 2008. "National Citizens' Technology Forum: Nanotechnologies and Human Enhancement." CNS-ASU Report R08-0002. http://cns.asu.edu/.

Hays, S., M.D. Cobb, and C.A Miller. 2013. "Public Attitudes towards Nanotechnology-Enabled Cognitive Enhancement in the United States." In *Nanotechnology, the Brain, and the Future,* Vol. 3 of *Yearbook of Nanotechnology in Society,* edited by S. Hays, J.R. Scott, C.A. Miller, and I. Bennett. New York: Springer.

Hibbing, J., and E. Theiss-Morse. 2002. *Stealth Democracy.* New York: Cambridge University Press.

Iyengar, S., and K. Hahn. 2009. "Red Media, Blue Media: Evidence of Ideological Selectivity in Media Use." *Journal of Communication* 59: 19-39.

Kleinman, D.L., J.A. Delborne, and A.A. Anderson. 2011. "Engaging Citizens: The High Cost of Citizen Participation in High Technology." *Public Understanding of Science* 20: 221-40.

Kosloff, S., J. Greenberg, T. Schamder, M. Dechesne, and D. Weise. 2010. "Smearing the Opposition: Implicit and Explicit Stigmatization of the 2008 U.S. Presidential Candidates and the Current U.S. President." *Journal of Experimental Psychology: General* 139, 3: 383-98.

Kunda, Z. 1990. "The Case for Motivated Reasoning." *Psychological Bulletin* 108, 3: 480-98.

Lupia, A. 1994. "Shortcuts versus Encyclopedias." *American Political Science Review* 88: 63-76.

Luskin, R., J. Fishkin, and K. Hahn. 2007. "Deliberation and Net Attitude Change." Paper presented at the ECPR General Conference, Pisa, Italy, 6-8 September.

Luskin, R., J. Fishkin, and R. Jowell. 2002. "Considered Opinions: Deliberative Polling in Britain." *British Journal of Political Science* 32: 455-87.

Macnaghten, P.M., M.B. Kearnes, and B. Wynne. 2005. "Nanotechnology, Governance, and Public Deliberation: What Role for the Social Sciences?" *Science Communication* 27: 268-91.

Macoubrie, J. 2003. "Conditions for Democracy." Paper presented at the annual meeting of the International Communication Association, San Diego, 23-27 May.

Mendelberg, T. 2002. "The Deliberative Citizen: Theory and Evidence." In *Research in Micropolitics: Political Decision Making Deliberation and Participation,* edited by M.X. Delli Carpini, L. Huddy, and R. Shapiro, 151-93. Greenwich, CT: JAI Press.

Merkle, D. 1996. "The National Issues Convention Deliberative Poll." *Public Opinion Quarterly* 60: 588-619.

Morrell, M. 2005. "Deliberation, Democratic Decision-Making, and Internal Political Efficacy." *Political Behavior* 27: 49-69.

Mutz, D. 2008. "Is Deliberative Democracy a Falsifiable Theory?" *Annual Review of Political Science* 11: 521-38.

National Science Board. 2010. *Science and Engineering Indicators 2010.* Washington, DC: National Science Foundation.

Nisbet, M.C., and D.A. Scheufele. 2009. "What's Next for Science Communication? Promising Directions and Lingering Distractions." *American Journal of Botany* 96: 1-12.

Nyhan, B., and J. Reifler. 2010. "When Corrections Fail: The Persistence of Political Misperceptions." *Political Behavior* 32, 2: 303-30.

Philbrick, M., and J. Barandiaran. 2009. "National Citizens' Technology Forum: Lessons for the Future." *Science and Public Policy* 36, 5: 335-47.

Pidgeon, N., B.H. Harthorn, K. Bryant, and T. Rogers-Hayden. 2009. "Deliberating the Risks of Nanotechnologies for Energy and Health Applications in the United States and United Kingdom." *Nature Nanotechnology* 4: 95-98.

Posner, R. 2005. "Smooth Sailing: Democracy Doesn't Need Deliberation Day." *Legal Affairs* 3, 1: 41-42.

Powell, M., and M. Colin. 2009. "Participatory Paradoxes: Facilitating Citizen Engagement in Science and Technology from the Top-Down?" *Bulletin of Science Technology and Society* 29: 325-42.

Powell, M., and D. Kleinman. 2008. "Building Citizen Capacities for Participation in Nanotechnology Decision-Making: The Democratic Virtues of the Consensus Conference Model." *Public Understanding of Science* 17, 3: 329-48.

Putnam, R. 2000. *Bowling Alone: The Collapse and Revival of American Community.* New York: Simon and Schuster.

Rawls, J. 1971. *A Theory of Justice.* Cambridge, MA: Harvard University Press.

Rowe, G.R., and L.J. Frewer. 2000. "Public Participation Methods: A Framework for Evaluation." *Science, Technology, and Human Values* 25: 3-29.

Rowe, G., R. Marsh, and L.J. Frewer. 2004. "Evaluation of a Deliberative Conference." *Science, Technology, and Human Values* 29: 88-121.

Ryfe, D. 2002. "The Practice of Deliberative Democracy: A Study of Sixteen Organizations." *Political Communication* 16: 359-78.

Sanders, L. 1997. "Against Deliberation." *Political Theory* 25: 347-76.

Sclove, R. 1995. *Democracy and Technology.* New York: Guilford Press.

Shapiro, I. 2002. "Optimal Deliberation?" *Journal of Political Philosophy* 10: 1-16.

Sprain, L., and J. Gastil. 2006. "What Does It Mean to Deliberate? An Interpretative Account of the Norms and Rules of Deliberation Expressed by Jurors." http://depts.washington.edu/.

Steenbergen, M.R., A. Bächtiger, M. Spörndli, and J. Steiner. 2003. "Measuring Political Deliberation: A Discourse Quality Index." *Comparative European Politics* 1: 21-48.

Stokes, S.C. 1998. "Pathologies of Deliberation." In *Deliberative Democracy,* edited by J. Elster, 123-39. Cambridge, UK: Cambridge University Press.

Sunstein, C.R. 2000. "Deliberative Trouble? Why Groups Go to Extremes." *Yale Law Journal* 110: 71-119.

–. 2003. "The Law of Group Polarization." In *Debating Deliberative Democracy*, edited by J. Fishkin and P. Laslett, 80-101. Malden, MA: Blackwell Publishing.

–. 2005. *Laws of Fear: Beyond the Precautionary Principle*. Cambridge, UK: Cambridge University Press.

Thompson, D.F. 2008. "Deliberative Democratic Theory and Empirical Political Science." *Annual Review of Political Science* 11: 497-520.

Wickson, F., M.D. Cobb, and P. Hamlett. 2012. "Review of Deliberative Processes: National Citizens Technology Forum – USA." In *Consumers and Nanotechnology: Deliberative Processes, Social Barriers, and Methodologies*, edited by H. Throne-Holst, E. Soto, P. Strandbakken, and G. Scholl. Singapore: Pan Stanford Publishing.

Wilsdon, J., B. Wynne, and J. Stilgoe. 2005. *The Public Value of Science: Or How to Ensure That Science Really Matters*. London: Demos.

Theorizing Deliberative Discourse

KIERAN O'DOHERTY

An aspect of public participation that has received relatively little attention is how particular recommendations or value statements are extracted from deliberative forums and integrated into some form of policy framework. That is, though proponents of public participation have argued for its importance, and different models of public consultation have been developed, the link between an arranged public engagement event and potential policy uptake remains relatively undertheorized. Some efforts in this regard have been suggested in the context of public deliberation on the topic of human tissue biobanking. In particular, previous work has focused on how the results of public participation can be meaningfully conceptualized (O'Doherty and Burgess 2009) and how public participation can be designed from the start such that the format of results is congruent with institutional practice and that the outputs of public deliberation thus would have a higher likelihood of being taken up in policy (O'Doherty and Hawkins 2010). In this chapter, I focus on one particular type of public engagement, public deliberation, with the aim of continuing the practical work of creating a conceptual and procedural foundation for linking the results of deliberation with policy uptake. In this context, the purpose of this chapter is to argue that there is a distinct structure to the discourse produced in deliberative forums and that this structure needs to be taken into consideration to produce a legitimate account of the outcomes of deliberation. I begin with a brief

overview of the goals of deliberative democracy, followed by some considerations of how these goals manifest in the context of emerging biotechnologies. I then introduce the notion of deliberative discourse and argue that, if
the nominal goals of deliberation have been achieved by a given group of
deliberants, then the discourse produced in that setting must necessarily be
characterized by certain key properties. I conclude by showing the relevance
of these considerations for the analysis of results of deliberative events.

The Goals of Deliberation

Public deliberation has a long and celebrated history, most often referred to
in the context of the city-states of ancient Greece and the town hall meetings of colonial New England (Delli Carpini, Lomax Cook, and Jacobs 2004).
Deliberative democracy also has a long existence as a subfield within political theory, which more recently has been supplemented by an increase in
efforts to design, implement, and evaluate deliberative forums. Deliberative
democracy has also come to feature as a valuable framework in the context
of recent trends in staging public consultations to inform science and technology governance (Hamlett 2003).

A significant amount of literature points to the desirability of involving
laypersons in policy and decision making in science, medicine, and technology (Collins and Evans 2002; Jasanoff 2003). Specifically, to be just and sustainable, collective decisions about new science and technology need to take
into account multiple value systems and consider potential impacts on different individuals and communities. This in turn requires input from lay
publics. Public and lay involvement in science and technology governance
can be promoted in a number of ways (Einsiedel 2002; Myskja 2007; Rowe
and Frewer 2005). One method of stimulating citizenship and addressing a
"democratic deficit" in policy development is through deliberative democracy exercises (Burgess and Tansey 2008; Fishkin and Laslett 2003; Gastil
and Levine 2005). Deliberative democracy is a process by which citizens are
given the opportunity to learn about a topic, engage others in debate about
the topic, and then come to a collective decision on what policy on the topic
should entail. Guidelines underlying deliberation focus on respectful engagement among participants, positions taken by participants that are justified and challenged by others, and conclusions that represent the deliberating
group's efforts to find common ground. The fundamental objective of deliberative democrats is to provide a process for citizen participation that enables discourse while avoiding manipulation (Dryzek 1990).

It is important to recognize features of deliberation that differentiate it from other forms of conversation. The definition of Chambers (2003, 309) is particularly useful here:

> Deliberation is debate and discussion aimed at producing reasonable, well-informed opinions in which participants are willing to revise preferences in light of discussion, new information, and claims made by fellow participants. Although consensus need not be the ultimate aim of deliberation, and participants are expected to pursue their interests, an overarching interest in the legitimacy of outcomes (understood as justification to all affected) ideally characterizes deliberation.

Gastil (2008, 8) similarly states that, "when people deliberate, they carefully examine a problem and arrive at a well-reasoned solution after a period of inclusive, respectful consideration of diverse points of view." Given these descriptions of deliberation, we should not expect the views or opinions expressed by deliberants to be constant over the course of deliberation. However, neither should we expect that the views offered by participants vary randomly. Rather, we might anticipate that there is a certain structure to the way in which expressed opinions evolve. In particular, we should expect that expressed opinions change from early to later stages of deliberation in the sense not only that opinions are technically more informed but also that they take into account a broader range of perspectives. But before elaborating this idea, I want to consider the particular case of deliberation on the topic of emerging biotechnologies.

The Case of Emerging Biotechnologies
The goals and principles of deliberative democracy outlined above are clearly applicable to a large range of contexts and topics. Within the range of issues to which deliberative democratic methods have been applied, the context of emerging biotechnologies is sufficiently distinct to warrant special attention to the characteristics of such forums. Examples of emerging biotechnologies about which public deliberation might be warranted include human tissue biobanks (large repositories of tissue samples used to facilitate medical research) and salmon genomics (Burgess, O'Doherty, and Secko 2008; O'Doherty, Burgess, and Secko 2010). Both examples do not (yet) feature prominently in public discourse, but they are nevertheless associated with difficult or controversial issues on a policy or regulatory level.

In particular, one can observe the following in the case of such emerging biotechnologies.

- Members of the public are likely to have little or no knowledge of the topic at the beginning of the deliberation.
- Consequently, members of the public might be less likely to express substantive opinions about many aspects of the topic prior to becoming engaged with the issues during deliberation.

For the outputs of deliberation to be considered legitimate, deliberants must demonstrate a minimum level of knowledge on the topic. Given the topic of an emerging biotechnology, if participants are to be lay citizens as argued above, then the design of the public engagement needs to include carefully constructed elements to inform deliberants about the topic under consideration.

There are two points to consider for such information provision. First, especially if the deliberation is about a controversial topic, the information provided to participants has to withstand scrutiny and charges of bias. In addition to ensuring that all processes are transparent, one effective way of addressing this problem is to present deliberants with information from different expert and stakeholder perspectives (MacLean and Burgess 2010). Second, individuals have different backgrounds and learning styles, and their capacities to assimilate and rates of assimilating information will differ. This can be best addressed by providing information through different modalities, such as written information, verbal presentations, online materials, and so on. However information provision is managed in a given deliberation, one would thus expect that in a successful deliberative forum, deliberants' individual and collective statements become increasingly informed as deliberants become more familiar with the topic. This is not to suggest a simplistic linear relationship between time and the degree to which statements are technically informed or accurate; rather, it is to point to a qualitative difference in the amount of information that, realistically, is likely to inform participants' statements at different points in time in the deliberation.

When dealing with a topic about which deliberants have little or no knowledge, they might also be less likely to express substantive opinions on it. Although individuals bring different base-line values to a deliberation, without sufficient information on the particular topic, it can be difficult to link these base-line values to the topic. It is difficult to generalize on this, but empirical evidence was gathered in at least one context in a deliberation on the issue of salmon genomics (Mackenzie and O'Doherty 2011). The deliberation was conducted among randomly selected residents of British Columbia who completed a survey on a range of items related to salmon genomics or contextual issues both before and after the event. The study showed a statistically significant shift in the number and strength of substantive opinions (defined here as choice of a non-neutral option on a survey item) that deliberants were able or willing to express at the end of the deliberation compared with the beginning. Although "salmon," on the West Coast of Canada, is a fairly politicized topic that is associated with strongly held views (O'Doherty, Burgess, and Secko 2010), this did not necessarily translate into individuals expressing strong views about salmon genomics, at least before deliberation. And, most significant for the arguments presented here, one would expect that this change in expressed opinions on the topic being discussed would be reflected in the discourse produced by deliberants over the course of deliberation.

Deliberation on emerging biotechnologies can have one further important difference from deliberation on other topics. One commonly cited characteristic of deliberation is that deliberants are expected to soften their views on a subject while opening themselves up to consider the views of others (e.g., Hamlett and Cobb 2006). The (often implicit) assumption here is thus that individuals have well-articulated and polarized views on the topic at hand. Indeed, one purpose of deliberation is to attempt to move individuals beyond their rigidly held views.[1] However, if the topic under consideration is an emerging biotechnology about which individual deliberants have little information, then this particular purpose of deliberation might not be applicable. As indicated above, the case of deliberation on salmon genomics illustrates that, for many issues related to the topic, individuals began to express opinions only as the deliberation progressed. In such cases, therefore, the purpose of deliberation can be reconceptualized as *producing substantive opinions*. This makes sense, for instance, if we want to use deliberation to inform policy on "future issues" (see Mackenzie and O'Doherty 2011) about which the general public is currently not aware but about which policy needs to be written.

Finally, some might argue that, even for topics about which individuals do not have much information, they are likely to express strong opinions owing to the emotional sensitivity of the topic or strongly held base-line values. In such cases, individuals can express strong opinions precisely because of the absence of good information. However, even in such cases, the goals of deliberation entail that deliberants are open to considering new information and the views of others and consequently willing to review their opinions. Thus, whether we are dealing with a topic in which individuals move from no opinions to substantive opinions, or from strong or rigid opinions to "softened" opinions, we should expect changes in the opinions expressed during deliberation.

The Structure of Deliberative Discourse

The discussion above argued that there is a distinct structure to the individual and collective opinions expressed over the course of a given deliberation. To make this point more clearly, I would like to introduce the notion of *deliberative discourse,* by which I simply mean the specific type of discourse produced by participants as they are engaged in deliberation. One can also conceptualize the deliberative discourse associated with a particular forum or event as the complete set of statements produced by deliberants in that setting.[2]

Thus, if a group of individuals has "successfully" deliberated (in the sense that they followed the processes and achieved the goals of deliberation as described above), then, by necessity, there must be particular structural properties that characterize the discourse arising from this setting. More specifically, when the participants are lay members of the public and the topic is an emerging biotechnology about which there is little public knowledge, the opinions expressed by deliberants can be expected to be increasingly informed and to encompass multiple views. Deliberative discourse under such conditions is thus likely to be characterized by a gradient evident in the opinions expressed by deliberants relative to

- information taken into account in the expression of a given opinion;
- the degree of strength or confidence of expressed opinions; and
- the range of values, perspectives, and opinions of others taken into account in the expression of individuals' opinions.

The first two properties listed here can probably be attributed to many situations in which an individual has some form of cognitive or affective engagement with a new area of knowledge. With increasing experience of engaging with a topic, individuals are likely to take into account increasing amounts of information and have increasing confidence in expressing an opinion. However, the third property is arguably fairly distinctive to deliberation. As explored above, a key aim of deliberation is for individual deliberants to feel safe in expressing their views while also being respectful of other deliberants expressing theirs. Over the course of deliberation, the views expressed by individuals should thus be expected to increasingly take into account values and interests that were not formative in the development of their original views. A corollary of this observation is that, overall, deliberative discourse is characterized by a kind of convergence toward not necessarily agreement, but certainly, mutual understanding.

Although some theorists uncompromisingly call for consensus as the only meaningful outcome of successful deliberation, the kind of convergence described above is well theorized and described by several theorists in more nuanced terms. Habermas (1996) uses the term "mutual understanding" to denote the ideal that individuals can hold and express views that encompass others' perspectives. According to this view, therefore, it is possible for individuals to show understanding of others' positions or beliefs, and accept them as valid, even when they disagree on the most appropriate action or decision in a given context. Drawing on this tradition, Niemeyer and Dryzek (2007, 500) explicitly identify metaconsensus and intersubjective rationality as goals of successful and authentic deliberation. They define *metaconsensus* as "agreement about the nature of the issue at hand, not necessarily on the actual outcome"; *intersubjective rationality*, they state, "results from deliberative procedure in which both agreement and disagreement are possible, but are constrained by a condition of consistency regarding the reasons that produce a particular decision," and "emerges when individuals who agree on preferences also concur on the relevant reasons, and vice versa for disagreement."

There is, thus, agreement among theorists in recognizing a certain degree of convergence of views or positions as central to the aims of successful deliberation. Theorists might disagree on whether they favour Niemeyer and Dryzek's criteria of metaconsensus and intersubjective rationality or Habermas's mutual understanding or whether they view only full consensus as a satisfactory outcome of deliberation. However, regardless of which view one might subscribe to, successful deliberation requires at least a degree of convergence of the expressed views of deliberants. Significantly, statements made during deliberation that are characterized by "mutual understanding" are arguably qualitatively different from statements made by individual deliberants from their perspectives *as individuals* (especially before they have been exposed to the views of other deliberants). Since it is a stated goal of deliberation to work toward mutual understanding among participants, the discourse of a successful deliberation should ideally demonstrate a shift from statements characterized primarily by the individual perspectives of participants to statements characterized more by consideration of plurality or collective interests. Taken as a whole, therefore, the deliberative discourse of a particular forum should manifest this asymmetrical shift in opinions expressed from the beginning of the deliberation to the end.

Finally, observations of the structure of deliberative discourse allow one to posit the construct of a *deliberative group opinion*. As already stated, opinions expressed by individual participants at the beginning of a deliberative event are likely to be manifestations of *individual opinion*. Such individual opinions can be understood, in the usual cognitive sense, to reflect some enduring value system internal to the individual or, in the discursive sense, to reflect some effort of the speaker at strategic self-representation (Potter 1996). However, if the ideals of deliberation as outlined above are achieved, then they should allow us to postulate a robust group opinion as a further outcome of successful deliberation. Whereas individual opinion can be aggregated to characterize the views of a group, such an aggregate construct nevertheless relies on independently formed opinions. In contrast, a deliberative group opinion formed as the result of successful deliberation is more than just an aggregate of individual opinions. Such a group opinion reflects statements that the group as a whole can agree to as a result of the respectful exchange of ideas, views, and opinions entailed in the process of deliberation. And though a clearly articulated group opinion can be reflective of consensus formed as a result of deliberation, this is not a necessity since it can also reflect clearly articulated persistent disagreements ("we agree to disagree about this"). Notably, the construct of a group opinion, as conceptualized here, is quite different from constructs such as the aggregate of large numbers of responses to political opinion or other surveys that are sourced independently from individuals who have not had meaningful mutual interactions with each other about the topic under consideration.

Implications for Analysis

I have argued that, given the stated goals of deliberation, the discourse produced in deliberative settings will inevitably exhibit a certain structure. These observations have important implications for the analysis of conversations among participants and the results of a deliberative forum along two possible dimensions. First, the observations above point toward potential criteria for the evaluation of the quality of deliberation in any given setting. That is, the discourse produced in a given forum could be evaluated in terms of the presence of certain structural elements to ascertain whether successful deliberation has in fact taken place. Second, assuming that successful deliberation has taken place, observations of the structure of the resultant discourse can help to improve how the results of a deliberative forum are presented and characterized. It is on this second point that I would like to focus more.

As discussed elsewhere (O'Doherty and Burgess 2009), precisely what constitutes the results of a deliberative forum is not self-evident. It is thus an analytical or theoretical decision how the results of deliberation are conceptualized and their achievement operationalized. For instance, it is a theoretical decision whether consensus is a required outcome of deliberation or not. Similarly, it is an analytical decision whether to define the results of a deliberation to be constituted by various statements made by participants, a summary produced by the moderator of the deliberation, or the product of some form of analysis prepared after the event. These decisions have consequences beyond the confines of the deliberation and, in particular, for how the results are understood and implemented in policy.

I want to focus here on how the structure of deliberative discourse outlined above might have consequences for the portrayal of the outcomes of a deliberation. In particular, if an analysis of the statements produced by participants over the course of a deliberation does not take into account the structural elements associated with the goals of deliberation, then the analysis is likely to come to unjustified conclusions. It is useful to consider the applications of standard qualitative methods, such as content or thematic analyses, to deliberative discourse. A thematic or content analysis, as applied to interview or focus group data, might be well suited to characterizing a comprehensive set of arguments around a certain topic. However, with such an approach, all arguments, preferences, and values expressed by participants over the course of deliberation are collapsed onto a single plane. In particular, without taking additional measures, typical content or thematic analyses would not be able to distinguish between statements made by deliberants (possibly at the beginning of the deliberation) based on individual opinions and made from relatively limited perspectives and statements (possibly made later on in the deliberation) that took into account the views of other deliberants and presented a more collective perspective. As outlined above, if the goals of deliberation have been operationalized appropriately in a given forum, then one would expect that, as deliberation proceeds, an increasing number of statements would take into account an increasing number of views and positions previously expressed by other participants; such statements would also increasingly take on the form of a collective or group opinion. Arguably, statements that take into account a larger range of views and positions, available in the deliberative forum as a whole, should also be accorded greater weight in any final analysis purporting to represent the results of the deliberation. This argument should not be taken to imply that certain views should be discounted because they do not explicitly acknowledge other points of view. For instance, if a participant provides the sole Muslim perspective in a deliberative forum and unwaveringly expresses her view on a subject without necessarily acknowledging other views, it does not mean that her views should be discounted in favour of other views shared by more participants. Rather, the point is that, given the goals of deliberation, there is likely to be variation in expression of the views of at least some individuals. In an analysis of the proceedings of a deliberation, this variation needs to be taken into account in a meaningful way. As is evident from ratification procedures conducted in some deliberative forums, at the end of a deliberative process, many individual participants are likely to endorse collective statements (or group opinions) rather than individual opinions that they might have expressed earlier on or before the deliberation. For analysis to be politically legitimate, different statements should not be given equal weight, but the commitment of participants should be followed as the expression of their views increasingly takes into account a broader and deeper appreciation of the issues.

A corollary of the foregoing is, thus, that analysis that characterizes the contents or outcomes of deliberation should ideally differentiate between statements made by deliberants as *individuals* and statements that are the result of several exchanges of views among participants and that therefore represent a collective statement (whether based on consensus, compromise, or a well-articulated polarized disagreement). Once again, in the context of presenting the results of a deliberation, there are strong reasons for placing greater emphasis on collective statements of participants, which are more likely to occur at the end of the deliberation than at the beginning.

Nevertheless, there are important reasons that this kind of weighting might not be applied in some kinds of analysis. In most deliberative forums, there is some element built into the process that operationalizes the results of the deliberation. This might be a report that is either produced by the deliberants themselves or with significant input from them, a presentation by deliberants given to some larger audience, or some form of summary proceeding by a non-participant such as a facilitator or note taker. Elsewhere I have termed the kind of output that reflects primarily the substantive input from deliberants, particularly when it is in the form of ratified collective statements, the deliberative output of a forum (O'Doherty and Burgess 2009). Deliberative output can be distinguished from what we might call analytical output, based on data from a deliberative event but involving substantive input from an analyst (e.g., collating all statements from participants over the course of the deliberation on a certain topic; producing an executive summary of the proceedings; or analyzing and evaluating the nature of interactions among participants). "Pure" deliberative outputs might have the advantage of increased political legitimacy, owing to their avoiding as much as possible analyst-mediated interpretation while also taking into account collective interests and a relatively higher amount of technical information. However, in reality, their utility to policy makers is likely to be limited without being complemented by appropriate analytical output, including appropriate uses of content analyses that apply equal weighting to statements made across the entire deliberation. In particular, analytical outputs by themselves do not demonstrate the conditions of their own validity. This requires additional analysis regarding issues such as whether information was adequately comprehended by participants, whether perspectives were actually shared, and whether dominant ideologies were reinforced or challenged.

An example of an appropriate, non-weighted use of thematic analysis of deliberative discourse is for the documentation and elaboration of reasoning underlying particular collective statements made by participants during deliberation. In the salmon genomics deliberation (O'Doherty, Burgess, and Secko 2010), one of the issues discussed at length by participants was the potential for transgenic salmon to be created and sold in supermarkets. Participants felt very strongly that, if this were to eventuate, they would want such GM salmon to be clearly labelled. This preference was explicitly stated and available in the deliberative outputs of the event, constituted by three ratified discussion reports based on presentations by participants. However, the deliberative output did not provide a clear understanding of the underlying motivation for the call for labelling. It was not clear from the collective statements alone why the participants felt so strongly about labelling. A thematic analysis conducted on the transcripts from the deliberation did, however, provide answers to this question (Nep and O'Doherty, forthcoming). In particular, in-depth analysis of all statements made by participants provided evidence that the call for labelling represented underlying factors such as strong distrust in the governance of GM foods directed toward researchers, the government, and biotechnology companies; a desire to be in control of the nature of one's sustenance; and a call for greater transparency. These insights are of evident value in considering policy on GM food labelling. Because they are not directly available from the deliberative outputs of the public engagement, additional in-depth (content) analysis of the deliberation was necessary to flesh out the recommendations of deliberants.

Conclusion

This chapter introduced the notion of deliberative discourse and argued for the recognition of its distinctive nature, specifically in the context of public engagement with emerging biotechnologies. I argued that, if public engagement successfully follows the principles of deliberative democracy, then there should be important systematic variations among participants' statements over the course of deliberation. Specifically, the discourse produced over the course of deliberation should be characterized by a certain gradient relative to particular statements being increasingly informed about the topic, increasingly aware of other (potentially opposing) points of view, and, therefore, increasingly collective in character. The notion of such a gradient should not be interpreted simplistically. As elaborated in detail in the field of discursive psychology (Edwards and Potter 1992), statements made by individuals need to be understood in the context of the local discursive purpose that the utterance serves. Moreover, though we might expect individuals to become more technically informed over the course of a deliberation, individual statements might still be more or less informed in a way that is not correlated positively with time. In other words, though I am suggesting that in an ideal situation, participants' discourse becomes more informed and intersubjective over the course of deliberation, I am not suggesting that it can necessarily be modelled by a simplistic linear relationship.

One conclusion that can be drawn from the observations made in this chapter is that some opinions are "better" than others, owing to being better informed and more inclusive of the views of other deliberants. Does this mean that analysis of the results of deliberation should only include statements that pass some threshold of being informed and inclusive of other opinions? Not necessarily. As pointed out in the example of deliberation on labelling of GM salmon, valuable information can be extracted from attempting to understand the arguments presented by individuals in the course of coming to some collective position. Thus, both ratified conclusions (deliberative output) and in-depth analysis of deliberation transcripts or other relevant data (analytic output) can produce insights that would be relevant to policy makers. However, it is important that how the data are extracted from deliberation events, how they are analyzed, and how the results or outcomes of a deliberation are characterized are appropriate to the particular knowledge claims being made. In the GM salmon case, for instance, it would be appropriate to characterize one of the results of the deliberation as a strong call for mandatory labelling because this was one of the conclusions ratified by a majority of deliberants at the end of the event. If some of the participants had been against mandatory labelling at the outset of deliberation and then changed their minds, then their original statements should not be given the same weight as the final collective statements when characterizing the outcomes of deliberation. However, if the purpose of an analysis is to track how individuals changed their minds on labelling GM foods as they were presented with new evidence or arguments from other deliberants, then, of course, all statements on the topic should be considered in the analysis.

The point of this chapter was a rather minor one: to argue that the act of deliberation produces a particular type of discourse (assuming that the stated aims of deliberation are met) and that analysis of this discourse needs to take this structure into account if it is to be legitimate. Several other questions can be raised, however. For instance, how might the structure of deliberative discourse differ between topics about which deliberants do not know much at the outset (e.g., emerging biotechnologies) and those with which they are more familiar (e.g., health care)? How might the structure differ between deliberative events that show evidence of having become polarized versus those that have not? Can observations about the structure of deliberative discourse be used to specify measurable criteria for evaluation of the deliberative quality of a public engagement? These are empirical questions that go beyond the theoretical focus of this chapter. However, their consideration relies on recognition of the structural properties of discourse produced in deliberation. To this end, I hope that the thoughts presented here provide a useful starting point for further inquiry.

NOTES

This research was supported by funding from Genome Canada through the Building a GE³LS Architecture and cGRASP projects. I am also grateful for helpful comments on an earlier draft of this chapter by members of the face-to-face research group at the W. Maurice Young Centre for Applied Ethics, University of British Columbia.

1 Sunstein (2002) argues that, because of the phenomenon of group polarization (i.e., the tendency for individuals within a deliberating group to reach more extreme conclusions compared with pre-deliberation opinions), one cannot assume that deliberants will soften their views. Sunstein also emphasizes the institutional parameters that seem to guard against group polarization, such as random selection of participants, unbiased presentation of information to deliberants, and availability of experts representing a diversity of views on the subject. Since the arguments presented in this chapter are about how to conduct analysis in cases in which deliberation of a relatively high quality has taken place, the presence of these elements in deliberative forums is assumed.

2 For the sake of simplicity, I confine myself here to consideration of deliberation in the sense of formally constituted forums, in which a group of citizens has been appropriately selected (usually randomly and/or using demographic stratification), informed, and professionally facilitated. Although some attempts have been made to extend the notion of deliberation to other settings, such as online discussion groups or other conversational settings (Danielson 2010; Gastil and Black 2008), these considerations go beyond the arguments of this chapter.

REFERENCES

Burgess, M.M., K.C. O'Doherty, and D.M. Secko. 2008. "Biobanking in BC: Enhancing Discussions of the Future of Personalized Medicine through Deliberative Public Engagement." *Personalized Medicine* 5, 3: 285-96.

Burgess, M.M., and J. Tansey. 2008. "Democratic Deficit and the Politics of 'Informed and Inclusive' Consultation." In *Hindsight and Foresight on Emerging Technologies*, edited by E. Einsiedel and R. Parker, 275-88. Vancouver: UBC Press.

Chambers, S. 2003. "Deliberative Democratic Theory." *Annual Review of Political Science* 6: 307-26.

Collins, H.M., and R. Evans. 2002. "The Third Wave of Science Studies: Studies of Expertise and Experience." *Social Studies of Science* 32, 2: 235-96.

Danielson, P. 2010. "A Collaborative Platform for Experiments in Ethics and Technology." In *Philosophy and Engineering: An Emerging Agenda*, edited by M. Davis et al., 239-52. New York: Springer.

Delli Carpini, M.X., F. Lomax Cook, and L.R. Jacobs. 2004. "Public Deliberation, Discursive Participation, and Citizen Engagement: A Review of the Empirical Literature." *Annual Review of Political Science* 7: 315-44.

Dryzek, J. 1990. *Discursive Democracy: Politics, Policy, and Political Science.* Cambridge, UK: Cambridge University Press.

Edwards, D., and J. Potter. 1992. *Discursive Psychology.* London: Sage.

Einsiedel, E.F. 2002. "Assessing a Controversial Medical Technology: Canadian Public Consultations on Xenotransplantation." *Public Understanding of Science* 11: 315-31.

Fishkin, J., and P. Laslett. 2003. *Debating Deliberative Democracy.* Malden, MA: Blackwell.

Gastil, J. 2008. *Political Communication and Deliberation.* Los Angeles: Sage.

Gastil, J., and L.W. Black. 2008. "Public Deliberation as the Organizing Principle of Political Communication Research." *Journal of Public Deliberation* 4, 1. http://services.bepress.com/.

Gastil, J., and P. Levine. 2005. *The Deliberative Democracy Handbook: Strategies for Effective Civic Engagement in the Twenty-First Century.* San Francisco: Jossey-Bass.

Habermas, J.R. 1996. *Between Facts and Norms: Contributions to a Discourse Theory of Law and Democracy.* Translated by W. Rehg. Cambridge, UK: Polity Press; Blackwell Publishers.

Hamlett, P.W. 2003. "Technology Theory and Deliberative Democracy." *Science, Technology, and Human Values* 28, 1: 112-40.

Hamlett, P.W., and M.D. Cobb. 2006. "Potential Solutions to Public Deliberation Problems: Structured Deliberations and Polarization Cascades." *Policy Studies Journal* 34, 4: 629-48.

Jasanoff, S. 2003. "Technologies of Humility: Citizen Participation in Governing Science." *Minerva* 41, 3: 223-44.

Mackenzie, M., and K.C. O'Doherty. 2011. "Deliberating 'Future Issues': Minipublics and Salmon Genomics." *Journal of Public Deliberation,* 7, 1. http://services.bepress.com/.

MacLean, S., and M.M. Burgess. 2010. "In the Public Interest: Assessing Expert and Stakeholder Influence in Public Deliberation about Biobanks." *Public Understanding of Science* 19, 4: 486-96.

Myskja, B. 2007. "Lay Expertise: Why Involve the Public in Biobank Governance?" *Genomics, Society, and Policy* 3, 1: 1-16.

Nep, S., and K.C. O'Doherty. Forthcoming. "Understanding Public Calls for Labeling of GM Foods: Analysis of Public Deliberation on GM Salmon." *Society and Natural Resources.*

Niemeyer, S., and J.S. Dryzek. 2007. "The Ends of Deliberation: Metaconsensus and Intersubjective Rationality as Deliberative Ideals." *Swiss Political Science Review* 13, 4: 497-526.

O'Doherty, K.C., and M.M. Burgess. 2009. "Engaging the Public on Biobanks: Outcomes of the BC Biobank Deliberation." *Public Health Genomics* 12, 4: 203-15.

O'Doherty, K.C., M.M. Burgess, and D.M. Secko. 2010. "Sequencing the Salmon Genome: A Deliberative Public Engagement." *Genomics, Society, and Policy* 6, 1: 16-33.

O'Doherty, K.C., and A. Hawkins. 2010. "Structuring Public Engagement for Effective Input in Policy Development on Human Tissue Biobanking." *Public Health Genomics* 13, 4: 197-206.

Potter, J. 1996. *Representing Reality: Discourse, Rhetoric, and Social Construction.* London: Sage.

Rowe, G., and L. Frewer. 2005. "A Typology of Public Engagement Mechanisms." *Science, Technology, and Human Values* 30, 2: 251-90.

Sunstein, C.R. 2002. "The Law of Group Polarization." *Journal of Political Philosophy* 10, 2: 175-95.

PART 4

INSTITUTIONAL CONTEXTS OF PUBLIC PARTICIPATION

Public Engagement and Knowledge Mobilization in Public and Private Policy

DAVID CASTLE

151-66

Academic researchers have been undertaking an increasing number of studies of public participation to gain insight into the public's awareness of, knowledge about, and interest in science and technology innovation. The public, or "publics" as some prefer, participate through a number of exercises, including deliberative democracy events, citizen juries, focus groups, and polling. Academics have a wide range of motivations for being interested in this research, and governments have an equally wide range of interests in seeing that the public can participate. Some events engage the public on topics such as risk and regulation without making an overt connection to any decision that needs to be taken by public sector decision makers. In other cases, participation of the public is consultative; it provides input that will be used in a decision. The distinction between engagement and consultation is often blurred, and it is not always possible to know how the outputs and outcomes of public participation are used. Because governments have an interest in maintaining their legitimacy as decision-making authorities, grounds for decisions are rarely attributed to studies or researchers. From the academics' point of view, demonstrating that their work has had uptake in and impact on public policy is therefore difficult. Private, non-profit organizations are not constrained in the same way as public sector institutions, providing another avenue for public participation to impact policy, albeit private policy in the first instance. Because of the growth of networked governance, however, a second occasion for uptake and impact can arise when private sector policies influence public sector decisions. Academic researchers might therefore aspire to work with private sector partners, particularly non-profit organizations, since policy relevance is easier to assure.

Public Engagement with Science and Technology Research and Development

The engagement of the public on aspects of science and technology innovation can be broken into four main phases: study design, data collection, analysis, and knowledge mobilization. Scholarly reflection on the methodology involved in the first three phases has a long tradition that is unmatched by discussion of the fourth phase. More attention is now being paid to knowledge mobilization, not only by researchers but also by funders of research who seek evidence of research impact. Consequently, there is a growing desire to ensure that research results are publicly disseminated, translated into formats that pique interest in and have relevance for specific groups, to achieve uptake by end users, and assessed in terms of their impact on policies and procedures of organizations. Since knowledge mobilization has come to mean all of these things, researchers are attending more closely to methodological considerations in knowledge mobilization and public engagement. This is done to include understandings of the purpose of an engagement exercise in relation to knowledge mobilization goals and the adaptation of study design, data collection, and analysis with respect to knowledge mobilization objectives.

Perhaps the most elusive of all aspects of demonstrating that effective knowledge mobilization has transpired is showing that there has been an impact on the policies and procedures of organizations. Knowledge mobilization that ends with knowledge transfer from researchers to other groups or institutions fulfills an important function. But even after knowledge transfer, the question of uptake by other groups and impact on policy remains. The outputs and outcomes of public engagement exercises can certainly satisfy intellectual curiosity and generate interesting results for reflection, discussion, and publication. Whether they are findings that cause people to do things differently, change procedures, or alter the course of policy development, and whether the impact is fleeting or enduring, are different matters.

Demonstrating the impact of public engagement work has different, and not always justifiable, motivations. First, there is the "research as activism" situation in which researchers blur or elide the distinction between academic research and social advocacy or activism. For such individuals, engaging the public to change society can be their paramount motivation. Some have argued that the role of the professor really ought to be research and professing on an area of expertise (Fish 2008).

Second, a "peer pressure" situation can occur in which researchers of a practical bent seek evidence that their academic research maintains connections to the world. They want to see the impacts of their work and might expect the same attitudes and expectations in others, regardless of whether all researchers share the same practical disposition. Indeed, given that there is room in the academy for researchers of different stripes, not all researchers will feel compelled to tilt toward practical impacts of their research just because some of their colleagues do, which is as it should be.

Third, there exists the "funder pressure" under which research results have to be reported in conjunction with some kind of impact statement. In Genome Canada reporting, for example, evidence of impact has to be reported as "socio-economic benefit." In the United Kingdom, the Research Assessment Exercise (RAE) is being revamped into a Research Evaluation Framework with up to one-quarter of the exercise being devoted to "impact." Interestingly, in each case, *ex ante* criteria that researchers could apply to their work are not available, leaving it to researchers to create their own impact standards, hoping for concurrence from project reviewers acting on behalf of funding agencies.

Fourth, there is the "public pressure" situation, which is considered at length in this chapter. In this scenario, there is a reasonable expectation among members of the public that their views will be solicited about major decisions planned and carried out on their behalf by government. In recent decades, there has been a growing expectation of stakeholder involvement in the governance of science and technology innovation and the regulation of products and services entering trade and commerce (Osseweijer 2006). Public opinion ought to count in governance decisions in a meaningful way, particularly when decisions will be made about research and commercialization involving new, known risks or when decisions will be made without knowing all of the risks (Barrett and Brunk 2007). But what does it mean to take public opinion into consideration?

Previous work (Castle and Culver 2006) defined an expanded role for academics engaging the public, in response to the oft-heard criticism that some academic work on public attitudes about new science and technology more closely resembles market research or government opinion polls than research. Whether this criticism has merit, or matters, is worth exploring.

Yet, such questions are distinct from thinking about an expanded and so-cially beneficial role for academic engagement of the public. The more ex-pansive role envisioned focuses on the problem of democratic disengagement generally and the feeling of powerlessness and disenfranchisement regarding new science and technology. The positive role for researchers undertaking public engagement exercises is not only to facilitate knowledge exchange among communities to improve understanding all around but also to clearly render problems and judgments into recommendations for action.

Various terms are used to describe interactions among researchers, gov-ernments, and the public, but some distinctions can be made. Generically speaking, all consultation and engagement are forms of public participation, and there are many different methods (e.g., focus groups and polling) by which people participate. In a more in-depth discussion in Castle and Culver (2006), varieties of public participation are distinguished in light of the goals of those organizing the event. Because the expanded role of researchers in engagement actually involves two separate components, a distinction can be drawn between public *engagement,* how one learns about public attitudes to science and technology innovation, and *consultation,* which refers to the inclusion of the public will in governance decisions. More precisely, the en-gagement of citizens by government, civil society, or other groups is the "push" of information to citizens via offline or online means such as town hall meetings and issue-based websites. Some engagement exercises can involve solicitation of citizens' views on issues related to the information provided, yet, it is not necessary for public engagement that any specific response be solicited. In contrast to engagement, public consultation in-volves not only the "push" of information noted above but also the "pull" of preferences from citizens. It is necessary for consultation that some actual decision is to be made, that citizens know a decision is to be made, and that their views are solicited and the consulting organization in fact receives them in a timely fashion relative to when the decision has to be made.

There is a temptation to think of engagement as nested in consultation, as if engagement were a logical or temporal precursor of consultation, but this is not necessarily the case. One can consult about a decision to be made without the common element of the push and pull of information and risk getting advice that is not contextualized or informed. If there is a norma-tive ranking between engagement and consultation, then the latter is to be preferred because it upholds democratic ideals, addresses democratic disengagement, ties consultation to actual decisions, and creates binding obligations to use the outputs of the consultation.

The public pressure reason noted above suggests a reversal of perspectives. Researchers want to have good processes and outcomes for their engagement and consultation exercises and naturally focus their attention on getting both. Members of the public attending engagement and consultation exercises have a variety of motivations, but their attending and participating are contingent on wanting to be heard (engagement) and having their views count in some meaningful way (consultation). Unfortunately, even for the most practical researcher who dutifully responds to funders' requests for evidence of impact, it is difficult to provide evidence, and not mere assurances, that the exercise is consultative. Part of the problem is that the paths for uptake of the results are subject to several pitfalls. The outputs arising from public engagement and consultation are typically published in academic, peer-reviewed journals in which they might have impacts on the work of other academics but can involve significant time lags or access barriers for intended users in government. Work done on consultancy might simply be required by statute without having an enthusiastic receptor in government; such work does not earn academic kudos but gives researchers the sense that their work has "policy relevance."

Promising to participants or research funders that a public engagement or consultation exercise will have an impact ultimately faces the same challenge. Government decisions are based on multiple lines of evidence collected from a wide array of sources. Documentable influence of government exists but most often occurs in the form of lobbying. Civil servants and elected officials have to be circumspect and inscrutable with respect to the use of other sources of information, and, in any event, they are not obliged to demonstrate *how* they have reached a decision; they are only required to be clear about *what* the decision is. Consequently, it is difficult, if not impossible, for the researcher to retrace all of the steps that lead from a policy or regulatory decision back to engagement of the public. The case in which a researcher has the opportunity to present research results to government officials, or has been involved directly in a government-supported exercise, will acquire as much evidence of impact as one can offer.

There are cases in which the situation is not so dire as the above suggests. For example, in 2007, the Council of Canadian Academies (CCA) undertook a study of the state of scientific knowledge on nanotechnology that would underpin regulatory perspectives (Expert Group on Nanotechnology of the Council of Canadian Academies 2008). The CCA formed an interdisciplinary expert panel in response to the request from the minister of health communicated through the minister of industry and took lay and expert opinion into account in the final report to Health Canada. As it turned out, some key recommendations of the Expert Group related to both Health and Environment Canada. One outcome was that Environment Canada decided to change the notification procedure for nanomaterials under the Canadian Environmental Protection Act. As was reported by the Canadian Broadcasting Corporation (2009), "the notice for nanomaterials will gather information that will be used towards the development of a regulatory framework and will target companies and institutions that manufactured or imported a total quantity greater than 1 kg of a nanomaterial during the 2008 calendar year." This change in the Environment Canada regulatory framework was supported by the public input received by the CCA panel that called for greater regulatory scrutiny of nanomaterials with respect to their human health and environmental effects (Einsiedel 2007). Overall, the CCA consultation with the public led directly to change in the policies and regulations of government and demonstrated the potential significance of knowledge mobilization.

Not all cases in which the public voice is heard on a subject have such definitive outcomes for the reasons given above, yet demonstration of impact is equally desired and perhaps as important. For example, the Canadian Food Inspection Agency (CFIA) has held, since 2002, statutorily required public participation with respect to plant-made products (PMPs), a class of biotechnology in which plants are engineered to create novel molecules of choice such as vaccines for human use. Because this involves novel technology, it is likely to raise questions about risk and ethics. There is an especially strong need for public engagement in cases in which the finished products are intended for human use (Castle and Dalgleish 2004). Academics have worked with the CFIA to understand, and ostensibly address, public concerns about the technology (Einsiedel 2007). Nevertheless, no significant regulatory decisions appear to have been based on the engagement exercises, and there is little research and development in PMPs in Canada at present, suggesting that, whatever the impact of engagement exercises in government, it is not visible in a regulated industry.

As intimated above, expectations need to be lowered about having documentable impact on public policy. Given the previously drawn distinctions among the goals of public participation events, most are engagement and not consultation events. There might still be direct, formal mechanisms by which government will consult, but, more often than not, academics working in universities or as consultants are undertaking engagement exercises. Because of this, drawing a connection between the engagement exercise and some policy outcome is difficult. This deserves qualification, however, because the policy outcomes typically of interest to academics lie in the sphere of public policy. That is, academic engagement of the public is focused on those who set government policy or are involved in establishing and monitoring regulatory frameworks. Perhaps they can no longer be derided as the inscrutable government mandarins of lore in the sense that interaction between researchers and decision makers in government is increasingly common. Yet, democratic governments in industrialized countries strive to retain their legitimacy by making decisions for which they remain accountable. Consequently, academics' desire to show direct influence on the policies and operations of government is always thwarted by governments' need to maintain legitimacy. In the end, demonstrating that one has had influence on government can be difficult if not impossible, especially in the short term.

So why do academics take on the challenge of proving the nearly unprovable by doing costly and time-consuming engagement exercises the results of which have no binding status in government policy and operations? Research as activism, peer pressure, funder pressure, and public pressure has been offered as partial and overlapping reasons. Perhaps another reason is historical – public participation has become a growth industry in the continuation of the responsible science movement born of the 1960s and 1970s environmental movements and overlapping concerns about the use of science in the nuclear era. A further motivation lies in trying to bring the results of research into governance: in the context of public participation, science-based regulatory agencies benefit from learning about non-scientific considerations that can affect the reception of new products and services. This is particularly the case if one avoids the charge of being an academic-based marketer by embracing a more expansive view of the role of public participation as a point of negotiation in discussions at the interface of science and society (European Commission 2007; MASIS Expert Group of the European Commission 2009), a stance reflected in the work of the Eurobarometer, for example (Gaskell et al. 2010).

Laudable as the efforts to make research count in the policies and operations of government might be, there are venues other than the public sector for uptake of the results of public engagement exercises. In addition to research uptake and consultancy to the for-profit private sector, there are opportunities to mobilize knowledge to the private, non-profit sector. Non-governmental organizations (NGOs) and quasi-non-governmental organizations (QUANGOs) are frequent users of public engagement exercises and partner with academics or use their results. There exists a less appreciated category, private, non-profit, professional organizations, which are similarly interested in the methods and outcomes of public engagement if they meet the need to inform their membership of public attitudes and trends. Private, for-profit and private, non-profit organizations are guided by policies and operating rules that might be reflected in corporate mission statements, letters patent, and internal policies and rules with varying degrees of public accessibility and accountability. Nevertheless, if researchers are interested in mobilizing knowledge, private institutions have governance frameworks that might be receptive to insights from public engagement exercises. Researchers have enjoyed consulting to the private, for-profit sector to the extent that universities now generally impose ceilings on the number of consulting days allowed and might have other requirements regarding how contacts are procured or managed within the university.

Private sector institutions of all stripes are generally at liberty to say who or what influences their decision making, even if they do not do so, and they are likewise at relative liberty compared with the public sector to impose on their hired consultants different degrees of contractually enforced confidentiality. Yet, it is arguable that the connection between public engagement exercises and knowledge mobilized by academics to the private sector affords a greater degree of traceability, even if one cannot predict with greater certainty at the outset of a relationship that the results of engaging the public will have greater impact. To underscore this point, the case study that follows discusses a situation in which the research team received research funds to study public as well as health-care professionals' attitudes toward new "health-related technologies." That term is used because the new products and services in question are nutritional genomics, about which there is some dispute whether it is health technology or lifestyle products and services (Castle et al. 2006; Castle and Ries 2009). The research team worked with a private, non-profit professional organization in developing the public engagement exercise and subsequently worked with the organization to develop policy-relevant publications that would inform the organization's membership and directly contribute to the development of policy on the topic.

Nutritional Genomics: A Case Study of Public Engagement and Knowledge Mobilization

The Advanced Foods and Materials Network (AFMNet; http://www. afmnet.ca) was created in 2003 to focus on food, food biotechnology, and bio-based technologies stemming from the food system. AFMNet is one of several Networks of Centres of Excellence and received nearly $40 million in funding to support the network, research, and knowledge mobilization. The network supported the project Social Issues in Nutritional Genomics: The Design of Appropriate Regulatory Systems and Issues of Public Representations and Understanding from 2006 to 2009. The focus of this research project was public awareness and understanding of nutritional genomics (nutrigenomics), the role of the media and hype, and the international comparative analysis of nutrigenomic regulations. The study of consumer and social issues concentrated on three main topics: (1) the adequacy of the Canadian regulatory framework regarding nutrigenetic tests, as it compares with those of other countries; (2) knowledge and attitudes of health-care professionals and members of the general public regarding nutritional genomics; and (3) media representation of the new technology.

To conduct research on these questions, a partnership was formed among researchers at the University of Ottawa, the University of Alberta, the Public Health Agency of Canada (PHAC), and the Dietitians of Canada, a non-profit organization of professional dietitians. PHAC has an interest in genomics as they relate to public health issues and has been a convenor of the group GRaPH-Int – Genome-Based Research and Population Health International Network (http://www.graphint.org/ver2/). Nutrigenomics, having attracted regulatory scrutiny in the United States, was also on PHAC's watch list, meaning that PHAC was interested in understanding how the technology was developing, which people had access to it, and where the need for regulation was arising. PHAC contracted Phoenix Strategic Consulting to assist in the development of the focus groups with the public and health-care practitioners, coordinate the focus groups, and report to PHAC (Phoenix Strategic Consulting 2007). The Dietitians of Canada consider nutrigenomics a potential area of expertise for their members since they have the nutritional background and can develop competency in genetics, whereas physicians receive very little training in both nutrition and genetics (Farrell 2009). The Dietitians of Canada were similarly involved from the start of the project and had been included as a partner in the original proposal.

The study design centred on focus groups and, to a lesser extent, a telephone survey. The focus groups, led by a professionally trained and prepared moderator, provided a way of exploring in-depth knowledge. Although focus groups might not generate quantitative data sets, and if they do they lack the numbers to create statistical power, they do allow for a nuanced dialogue about a topic. In the case of science and technology innovation and the potential entry of new products and services into the market, focus groups are particularly useful for scoping the core issues that might arise. The novelty of nutrigenomics presents this kind of opportunity for discussion about the potential benefits and risks associated with the technology since nutrigenomics is not in widespread use. Consequently, the focus groups involved members of the public and health-care professionals, but in neither case was prior knowledge about nutrigenomics assumed or preferred.

Twelve focus groups were convened in November 2007 to assess the knowledge of nutrigenomics and attitudes toward nutritional genomics of nearly 100 randomly selected members of the public. The focus groups were held in five Canadian cities (Halifax, Montreal, Toronto, Edmonton, and Vancouver), and each session had approximately eight participants of reasonable demographic diversity, considering the overall small sample size. At least half of the participants used the Internet to search for health-related information, an important consideration in light of direct-to-consumer marketing of nutrigenetic tests, and a similar proportion of participants had children, an important consideration in light of the relational relevance of genetics. With respect to the health-care professionals, two focus groups were held in Toronto and two in Vancouver, totalling twenty-five participants. Physicians, pharmacists, dietitians, nutritionists, and naturopaths were selected to represent various practice settings. In each case, the focus groups lasted approximately two hours, consisting of a brief questionnaire in which, among other questions, participants were asked to explain what they knew of, and thought about, "nutrigenomics" and "personalized nutrition."

All respondents were given a brief explanation of the term "nutritional genomics" to establish a common starting point. It was described as

> a new, developing science that studies the way our genetic make-up affects how our bodies respond to what we eat and drink, where testing is often done by taking a sample of saliva, examining people's genetic make-up, and then providing them with nutrition- and diet-related information – a kind of personalized nutrition plan – to potentially help them improve their health and reduce their risk of certain diseases.

Questions about the importance of food in relation to health and knowledge about genes were asked to set the context before turning toward awareness and initial perceptions of nutrigenomics, reactions to a media article, a mock website selling nutrigenomic tests and services, and a summary document describing the current state of nutrigenomics. Some months later, a nation-wide telephone survey of 857 randomly selected adults used questions similar to those discussed in the focus groups.

Analysis of the results of the focus groups is reported by University of Ottawa and University of Alberta researchers in two peer-reviewed publications, one reporting on the public focus groups (Castle et al. 2010), the other on the health-care professionals' awareness of, and attitudes toward, nutrigenomics (Weir et al. 2010). In both cases, generally low levels of awareness were reported, with interest in nutrigenomics being moderated by concerns about the accuracy and validity of the tests. Phoenix Strategic Consulting (2007) reported similar results in its report to PHAC. The Dietitians of Canada received these papers and results, and two other peer-reviewed short articles were written for the dietitians (Morin and Castle 2009a, 2009b). Three other articles in the Practice-Based Evidence in Nutrition (PEN) series targeted the Dietitians of Canada membership directly and were developed in close consultation with the director of public affairs for the Dietitians of Canada (Morin et al. 2009, 2010; Weir et al. 2009).

Collaboration between the university-based research team and PHAC and the Dietitians of Canada, one that was initiated when the application for funding was written, was fruitful. This is an important factor in the success of the knowledge mobilization, as both institutions were engaged and had made an up-front commitment of staff time and accessibility to the project. They acted as arm's-length co-investigators, in the sense that they had direct input into development of the focus group guides but were neither primarily responsible, nor in a position to make overriding decisions, regarding the content and implementation. Participation in the early stages of the research project helped to create the conditions under which meaningful knowledge mobilization could happen once the research results were analyzed, a fact that might be viewed by academics as "conditioning the receptors" but might be viewed by PHAC and the Dietitians of Canada as "conditioning the researchers."

PHAC was explicit from the outset that genomics, and, in particular, potential public health applications or implications of nutrigenomics, were a priority for the agency. The research undertaken helped PHAC in its work to understand the potential issues and opportunities that could arise with different kinds of access to genetic testing and the role of health-care practitioners as potential providers of tests or information if they are asked to interpret the tests. The Dietitians of Canada were similarly explicit about their goals, recognizing that their members might be asked by the public to explain nutrigenomics, access tests, and interpret and act on results of tests. The Dietitians of Canada were also interested in the potential regulatory oversight of nutrigenomics that would affect members' ability to access tests or constrain the advice given based on test results. Both goals were clearly linked to awareness that the research would contribute directly to the identification of potential new opportunities for dietitians. New opportunities were associated with the need to ensure that dietitians receive training and are accredited appropriately. The focus group research, coupled with regulatory and media analysis, provided a basis on which the professional organization could make decisions about the extent and kind of information that its members receive, and that information was developed and disseminated with the oversight of the Dietitians of Canada.

Implications for Governance and Reconceiving the Role of the Public

The nutrigenomics case study demonstrates how researchers engaging the public about new science and technology can mobilize research results and have some immediate and documentable impact. The two receptors, PHAC and the Dietitians of Canada, have clearly discernible interests in the public engagement exercise. PHAC is an agency of Health Canada, a science-based regulator, and therefore benefits from learning about non-scientific considerations that would affect the reception of new products and services. The Dietitians of Canada is a non-profit professional organization, and its interest lies in both the methods and the outcomes of public engagement if they meet the need to inform its membership of public attitudes and trends.

Returning to the motivations for engaging the public, researchers involved in the project were not pursuing research as activism because none of them had commitments to the field of nutrigenomics, to associated research and development of products and services, to PHAC, or to the Dietitians of Canada. Peer pressure to demonstrate impact exists in the community of researchers who work on the social science aspects of science and technology innovation; however, insofar as the research team has published, and will continue to publish, theoretical papers in addition to empirical work, peer pressure is not a strong motivator. There is a clear motivation arising from funder pressure because the social sciences work of AFMNet, a predominantly science-based research network, is supposed to be integrated with natural science. Furthermore, the Networks of Centres of Excellence program creates networks that are supposed to translate academic research into products and services that benefit the well-being of Canadians through a variety of mechanisms, including private-public partnerships. Social science research, including public engagement, must ultimately serve this objective for the funds to be received and spent in good faith. Finally, though the project was guided by the view that science and technology innovation needs to be socially accountable, an explicit motivation arising from public pressure did not exist. Instead, public pressure was understood as the legitimate demand of citizens that their governments maintain policies and regulations that pursue a democratically decided norm of acceptable risk.

This last point raises an issue for the future of public engagement and the expectations that researchers might have for uptake and impact of their work. The above discussion on the relationship between engagement and consultation and democratic ideals resonates with a particular view of government and its relationship to the advice that it receives. Hierarchical government organization consisting of vertically managed line departments treats engagement and participation in predictable but unsatisfactory ways. There remains a tension in hierarchically organized government where inputs might be collected legitimately (deLeon and Varda 2009), but the actual decisions transpire among actors in a small and isolated network of knowledge brokers representing expert communities (Greenaway, Salter, and Hart 2007), with actors who can move fluidly in and among policy networks (Coleman and Perl 1999). Hence the above urging to distinguish engagement from consultation, to inform the public of the difference when involving it, and to make appropriate and transparent use of public participation. Some go further to urge that citizens act as "representatives" in public participation exercises, implying that democratic ideals would be more completely satisfied if different relationships between citizens and governments were configured (Warren 2008).

Further complicating matters is that some believe that a shift from government to governance is occurring in which a horizontal steering of actors to policy objectives (Rhodes 1996) replaces top-down directives – for example, "nudge economics" replace *dirigiste* economics. In the context of science and technology innovation, the triple helix model of government-academic-industry interactions involves the blurring of responsibilities and activities, where one of the three does the work of the others. These mixed networks can blur traditional lines of accountability as it becomes less clear to whom specific knowledge claims are to be attributed. Governance of innovation is especially challenging because it is a paradigm case of networked governance – no one can "pick the winners," least of all government. A wide array of actors from the three strands of the triple helix must be coordinated in a network that enables innovation by creating conditions under which knowledge flows yield maximum social benefits.

The implications for consultation and engagement are clear. Consultation in hierarchical government presumes that there is an authority, a pending decision, and a statutory obligation to collect and perhaps use outputs and outcomes of public consultation. These criteria can become less relevant if the hierarchical structure of government that supports them transitions to networked governance. The fate of uptake and impact of engagement outputs and outcomes is similarly compromised, particularly since governments can be quite selective without being transparent about the advice to which they attend. At first glance, the transition from government to governance might seem to jeopardize the demand for the public accountability of science and the responsibility of scientists in their research and of developers in their creation of new technology. Perhaps it does, in a conventional manner of thinking about the uptake and impact of public engagement. At the same time, if effective governance of science and technology innovation (i.e., wealth- and welfare-creating governance) is networked, then pathways to non-scientific considerations open up. Hierarchically organized, science-based departments of government rule out explicit use of non-scientific considerations in decision making. The private sector is not so encumbered, which explains why private, non-profit organizations are worth considering if one wishes to have uptake and impact from public engagement.

REFERENCES

Barrett, K., and C.G. Brunk. 2007. "A Precautionary Framework for Biotechnology." In *Genetically Engineered Crops: Interim Policies, Uncertain Legislation,* edited by I.E.P. Taylor, 133-52. New York: Haworth Press.

Canadian Broadcasting Corporation. 2009. "Ottawa Eyeing Nanotech Safety." 29 January. http://www.cbc.ca/news/.

Castle, D., C. Cline, A.S. Daar, C. Tsamis, and P.A. Singer. 2006. *Science, Society, and the Supermarket: The Opportunities and Ethical Challenges of Nutritional Genomics.* Hoboken, NJ: John Wiley and Sons.

Castle, D., and K. Culver. 2006. "Public Engagement, Public Consultation, Innovation, and the Market." *Integrated Assessment Journal* 6: 137-52.

Castle, D., and J. Dalgleish. 2004. "Cultivating Fertile Ground for Plant-Derived Vaccines." *Vaccine* 23: 1881-85.

Castle, D., and N.M. Ries. 2009. *Nutrition and Genomics: Issues of Ethics, Law, Regulation, and Communication.* Burlington, MA: Elsevier Press.

Castle, D., M. Weir, K. Morin, and N.M. Ries. 2010. "Public Knowledge, Awareness, and Perceptions of Nutrigenomics: The Canadian Perspective." *Teknoscienze: Agro FOOD Industry High-Tech* 21: 14-17.

Coleman, W.D., and A. Perl. 1999. "Internationalized Policy Environments and Policy Network Analysis." *Political Studies* 47: 691-709.

deLeon, P., and D.M. Varda. 2009. "Toward a Theory of Collaborative Policy Networks: Identifying Structural Tendencies." *Policy Studies Journal* 37: 59-74.

Einsiedel, E. 2007. *Pharming the Future: A Citizens' Panel Perspective on Plant Molecular Farming.* http://www.ucalgary.ca/pharmingthefuture/.

European Commission. 2007. *Taking European Knowledge Society Seriously: Report of the Expert Group on Science and Governance to the Science, Economy, and Society Directorate, Directorate-General for Research, European Commission.* Luxembourg: European Commission.

Expert Group on Nanotechnology of the Council of Canadian Academies. 2008. *Small Is Different: A Science Perspective on the Regulatory Challenges of the Nanoscale.* Ottawa: Council of Canadian Academies.

Farrell, J. 2009. "Health Care Provider Capacity in Nutritional Genomics: A Canadian Case Study." In *Nutrition and Genomics: Issues of Ethics, Law, Regulation, and Communication,* edited by D. Castle and N.M. Ries, 139-60. Burlington, MA: Elsevier Press.

Fish, S. 2008. *Save the World on Your Own Time.* Oxford: Oxford University Press.

Gaskell, G., et al. 2010. "Europeans and Biotechnology in 2010: Winds of Change?" Eurobarometer 73.1 on the Life Sciences and Biotechnology. Brussels: European Commission. http://ec.europa.eu/public_opinion/.

Greenaway, J., B. Salter, and S. Hart. 2007. "How Policy Networks Can Damage Democratic Health: A Case Study in the Government of Governance." *Public Administration* 85: 717-38.

MASIS Expert Group of the European Commission. 2009. *Challenging Futures of Science in Society: Emerging Trends and Cutting-Edge Issues.* Brussels: European Commission.

Morin, K., and D. Castle. 2009a. "Nutritional Genomics: Are Canadian Consumers Interested? Should Dietitians Be?" *Dietitians of Canada Current Issues,* April: 1-4.

—. 2009b. "Nutritional Knowledge and Perception of Nutritional Genomics among Canadian Healthcare Professionals." *Dietitians of Canada Current Issues,* June: 1-3.

Morin, K., M. Weir, N.M. Ries, and D. Castle. 2009. "Issues Related to the Delivery of Services Based on Nutritional Genomics." Dietitians of Canada Practice-Based Evidence in Nutrition (PEN). http://www.dieteticsatwork.com/.

—. 2010. "Regulatory Issues in Nutritional Genomics." Dietitians of Canada Practice-Based Evidence in Nutrition (PEN). http://www.dieteticsatwork.com/.

Osseweijer, P. 2006. "A Short History of Talking Biotech: Fifteen Years of Iterative Action Research in Institutionalizing Scientists' Engagement in Public Communication." Vrije Universiteit Amsterdam.

Phoenix Strategic Consulting. 2007. *Final Report: Exploring How Canadians Understand Nutrigenomics: Results of Focus Groups with Members of General Public and Health Care Professionals.* Ottawa: Phoenix Strategic Consulting.

Rhodes, R.A.W. 1996. "The New Governance: Governing without Government." *Political Studies* 44: 652-67.

Warren, M. 2008. "Citizens as Representatives." In *Designing Deliberative Democracy,* edited by M. Warren and H. Pearse, 50-69. Cambridge, UK: Cambridge University Press.

Weir, M., K. Morin, N. Ries, and D. Castle. 2010. "Canadian Health Care Professionals' Knowledge, Attitudes, and Perceptions of Nutritional Genomics." *British Journal of Nutrition* 104, 8: 1112-9.

–. "Ethical Issues in Nutritional Genomics." Dietitians of Canada Practice-Based Evidence in Nutrition (PEN). http://www.dieteticsatwork.com/.

From Public Engagement to Public Policy
Competing Stakeholders and the Path to Law Reform

DAVID WEISBROT

In recent times, there has been growing popular frustration in many liberal Western societies with the inability of parliaments and other representative democratic institutions to develop and implement social policy. Among other things, this has been caused by the polarization and divisiveness of increasingly partisan politics, the lack of human and financial resources – and parliamentary time – allocated to policy development, and the stultifying effects of the modern "twenty-four-hour news cycle." This cycle, with its voracious need for "news," and especially controversy, makes politicians and senior public servants reluctant to articulate and explore ideas for reform without having endlessly workshopped them in advance and considered the polling. The current political and media culture in Western societies does not encourage (or even forgive) political figures operating as public intellectuals, floating ideas for public consideration and debate, and then refining or abandoning them as appropriate.

In these circumstances, institutional law reform bodies – usually called "law reform commissions" – can play a very important role both in developing intelligent public policy and in directly engaging the community and providing the space for public participation in the policy development process.

In this chapter, I explore the role of the Australian Law Reform Commission (ALRC), in particular, in managing large-scale public inquiries, over the past decade, into the protection of human genetic information and Australian privacy law and practice. Both inquiries involved the need to develop new strategies for dealing with the impact of rapidly changing science and technology and the consequent ethical, legal, social, and technical implications. Given the novelty and dynamic nature of these areas, governments could not hope to find an "off the shelf" solution, requiring the ALRC to undertake a sophisticated program of interdisciplinary research.

Similarly, given the powerful and conflicting interests involved, the ALRC had to carefully manage the process to ensure that all stakeholders thought that they had been given a meaningful opportunity to present their own case for reform. For example, insurers and credit-rating agencies have a strong commercial interest in accessing as much (relevant) information as possible about individuals to carry out their underwriting and risk management functions, whereas individuals and privacy advocates want to ensure that such access is governed by strict limits and protections.

Institutional Law Reform

Institutional law reform commissions made their first appearance in 1965 in the United Kingdom (England, Wales, Scotland, and Northern Ireland) and Ireland and then rapidly spread throughout the Commonwealth world, including the Australian states and territories; New Zealand and the Pacific Islands (Papua New Guinea, Solomon Islands, Fiji, and Samoa); Canada (at both the federal and the provincial levels); Hong Kong and South Asia (India, Sri Lanka, Pakistan, and Bangladesh); the Caribbean (Trinidad and Tobago and Jamaica); and Eastern and Southern Africa (South Africa, Namibia, Malawi, Lesotho, Kenya, Uganda, Tanzania, Zaire, and Zimbabwe) (Tilbury 2005).

Creation of the ALRC by statute in 1973[1] coincided with a broader societal change in thinking about the role of law and law reform in driving progressive social change. Until the mid-1960s, work by state law reform committees was mainly focused on aspects of "black letter law," seen to be "technical" and thus the province of judges and lawyers. However, the mood of the community had begun to shift in the 1960s, demanding more opportunities for direct participation in the democratic process and greater accountability and transparency of public institutions.

The establishment of institutional law reform commissions in the 1960s and 1970s fit snugly within the "modernist" project of that era, which featured

- the strong faith in progress through specialist expertise and technocratic solutions;
- the view of law as a neutral technology, providing solutions as applicable to the problems of Indigenous communities and the Third World as to those of advanced, industrialized societies (Burg 1977; Merryman 1977; Trubek and Galanter 1974);[2]
- the belief in the socially transformative power of "big law" through omnibus legislation and high-powered, public interest, test case litigation; and
- the belief that government can, and should, play a central organizing role in such activities.

These sentiments are reflected in the statutory charters common to almost all institutional law reform agencies,[3] which generally call for the body to "systematically develop and reform law" by

- adapting the law to current conditions and needs;
- removing defects and obsolete or unnecessary laws;
- simplifying the law;
- adopting new or more effective methods of administering the law and dispensing justice;
- improving access to justice; and
- consolidating and harmonizing laws while having regard to individual human rights, the International Covenant on Civil and Political Rights, international obligations, and the costs of gaining access to and dispensing justice (Australian Law Reform Commission 1996).

Thus, it is no surprise that, during the same period in the mid-1970s that the ALRC was established, Australia also saw

- the creation of the Human Rights and Equal Opportunity Commission (HREOC, now called the Australian Human Rights Commission) and the Commonwealth Ombudsman's Office;
- the establishment of the Federal Court of Australia, the Family Court of Australia, the Administrative Appeals Tribunal, and other federal merits review tribunals promoting administrative review of official decision making;
- the recognition of the compulsory jurisdiction of the International Court of Justice as well as ratification of a range of major human rights and anti-discrimination treaties; and

- the assumption of federal responsibility for legal aid and a massive increase in funding and accessibility of legal services.

At the grassroots level, this expansion of legal aid was mirrored by the advent of the Aboriginal Legal Services and Community Legal Centres movements, which were

> born of the social activism and politicisation of the late 1960s and 1970s. The legal centres ... consciously developed as an alternative to the existing models of legal services delivery, embracing the political ideology and strategy of the welfare movement: a commitment to grass-roots level activity, community control, empowering the recipient, de-professionalization, assertion of rights, demystification and free access to services. (Weisbrot 1990, 246)

Similarly, there was a growing sentiment that Australian laws and legal institutions should be "relevant," reflecting contemporary conditions and community attitudes. Unlike the British law reform commissions, which stuck to matters of "lawyers' law," the Australian government asked the ALRC to work on matters with a strong social policy emphasis, leading to reports in its first decade of existence on complaints against police; police powers; alcohol, drugs, and driving; insolvency and bankruptcy; human tissue transplants; privacy; defamation; sentencing of federal offenders; insurance contracts and agents; child welfare; and the recognition and application of Aboriginal customary law.

Social Change and the "New Principle" of Modern Law Reform

Observing the creation and spread of institutional law reform agencies, the noted expert on government and public administration, Professor Geoffrey Sawer (1970, 183), wrote that this development in Australia and overseas reflected "the qualitatively new principle ... that the whole body of the law stood potentially in need of reform, and that there should be a standing body of appropriate professional experts to consider reforms continuously." Increasing questioning of the established institutions of that time included a lack of faith in other alternatives for bringing about legal change. This skepticism extended to

> the judges, who were generally unwilling to reform the law by court decision and were still dominated by the literal rule of statutory interpretation,

and parliaments, which were no longer willing to enact law reform by copying Imperial legislation, and the law was seen as not keeping up with technology and changes in social values; and yet there was an optimistic belief that state-sponsored activity could cure social problems. (Handford 1999, 506-7)

Senior members of the judiciary themselves expressed reservations about whether the courts were the best vehicles for systematic and comprehensive law reform. In 1979, Justice (and later Chief Justice) Sir Anthony Mason noted that the responsibility of the High Court of Australia is "to decide cases by applying the law to the facts as found" and that the court's techniques and procedures were adapted to that responsibility and not to legislating or engaging in law reform activities.[4] Several years earlier, Sir Anthony had raised the intriguing possibility of law reform commissions being given delegated legislative authority, subject to the power of disallowance by either House of Parliament (Mason 1975).

For Sawer (1970, 185), this "new principle of law reform" should be embodied in a commission with four distinguishing characteristics. It should be permanent, full time, independent, and authoritative. The call for permanent, full-time commissions was a reaction to the poorly resourced, ad hoc committees that traditionally operated in the law reform space. Permanent bodies would be in a better position to command and marshal financial resources, recruit and retain specialist staff, develop systems, and maintain corporate memory.

Independence is the *sine qua non* of institutional law reform. Although law reform agencies are entirely, or mainly, funded by government – and must remain accountable for the proper use of public funds – it is essential that they maintain an intellectual independence from government in the same manner that courts and ombudsman's offices do. Most statutes establishing law reform commissions expressly refer to the independent nature of the institution, but this is at least as much a cultural issue as a technical legal one.[5] Law reformers must jealously guard their ability to operate as they see fit and, ultimately, to deliver reports and advice without fear or favour, and they must be especially careful never to be put in a position in which they might be seen to be identified with partisan politics or special interests.

It also means that the institutional culture must be sufficiently robust to weather strong criticism at times. Few other institutions in our society are as accustomed as law reform commissions to sharing openly their work in progress. This invariably means having some preliminary views and "trial balloons" shot down along the way – with some bruised egos, perhaps, but in the ultimate interests of better policy making. As noted above, our immature media and political cultures now make it very difficult for governments to formulate and refine policy in this way: a politician who floats ideas or is persuaded by someone else tends to be roasted for "inconsistency" or "flip-flopping" rather than being applauded for creativity or flexibility. Institutional law reform provides an outlet for this more considered policy development process, but because of its non-partisan nature, there are fewer reasons for cynical point-scoring, and thus, law reformers tend to attract much less heat than politicians.

Apart from the inherent importance of intellectual integrity, there are practical reasons for ensuring that law reform commissions are, and are seen to be, independent. Access to confidential information is often valuable to the policy-making process – whether this is commercial-in-confidence material from industry, classified information from government, or highly sensitive personal stories from individuals. Stakeholders must also feel comfortable coming forward with views that are critical of established authority. Confidential submissions are regularly made to law reform commissions and contain sensitive information that would never be revealed to "the government."

Similarly, an independent law reform commission can generate and harness an extraordinary volunteer effort in the course of its work to supplement in-house research and expertise. For example, it is now standard operating procedure for the ALRC to establish a broadly based, expert Advisory Committee to assist in the development of all its inquiries. The ALRC solicits the involvement, on a pro bono basis, of the acknowledged leaders in their respective fields – and such invitations are rarely refused. No doubt this is attributable to the opportunity to influence policy and practice in an area that such individuals feel passionate about and the real and perceived independence of the ALRC. No judge, professional, scholar, community leader, or business executive would volunteer services to a body seen to be under the influence or control of the government.

Further Attributes of a Successful "Postmodern" Law Reform

To Sawer's (1970) classic list, I would add four characteristics essential for a contemporary law reform commission. Apart from the attributes specified by Sawer, a twenty-first-century law reform body must also be generalist, interdisciplinary, consultative, and implementation minded. To some extent, these additional attributes simply elaborate aspects of Sawer's fundamentals. However, these additions also reflect the changing political and social climate in which the law reform process currently operates.

Under what might be called this postmodern sensibility,

- there are doubts about the "traditional certainties" and a questioning of traditional authority, including public institutions;
- there is a greater appreciation of the complexity of social institutions and problems, including the fact that there might be intractable competing interests for which no easy compromise or consensus solution is possible;
- power is seen to be much more diffused and not entirely invested in the formal organs of government;
- governments are not necessarily seen as central to the solution of all social problems, with a preference for private sector or community-based strategies in at least some circumstances;
- there is greater reluctance to see all disputes as "legal"; and
- there is an increasing clamour for mechanisms that enhance opportunities for genuine public participation in civil society and public policy making.

As a consequence, there is greater skepticism of blockbuster legal solutions, with a preference for more textured, holistic strategies that place greater emphasis on process, education, communication, and allocation of responsibility and authority to multiple stakeholders.

This requires a clear sense of both the possibilities and the limitations of law reform. If it were ever the case, the old conundrum of "do you recommend the ideal solution or the pragmatic one?" now has little resonance. In the current political era, public sector funding is highly contested, and deregulation and privatization have dispersed power and responsibility. These days a law reform commission must think very carefully about any recommendation that entails significantly increased public expenditure: far more than was the case some years ago, it must spell out the precise amounts, offsets, and cost-benefit analyses involved.

In an earlier era, the centrepiece of any significant law reform effort was the recommendation of a new piece of legislation or a major amendment to existing law. However, in a more complex environment in which authority is much more diffused, modern law reform efforts are likely to involve a mix of strategies and approaches, including legislation and subordinate regulations; new dispute resolution options; official standards, guidelines, and codes of practice; voluntary industry codes; education and training programs; better coordination of governmental and intergovernmental programs; personal and professional ethics; institutional restraints; peer review and pressure; oversight by public funding authorities and professional associations; supervision by public regulatory and complaints-handling authorities; occasional media scrutiny and exposé; private interests; and market pressures.

The very existence of an open inquiry is often enough to stimulate introspection, internal review, or reform by the institutions under scrutiny. Social scientists have described the "Hawthorne effect" (and natural scientists, the "Heisenberg effect"), by which the observation of an activity itself has an impact on that activity. This "spotlight" phenomenon has been apparent to the ALRC in recent inquiries. For example, the inquiry into the protection of human genetic information spurred the peak life insurance body in Australia, the Insurance and Financial Services Association (IFSA), to commission empirical research, review its policies and practices, and develop educational literature for consumers and salespeople. Employers and trade unions, which had not considered these issues at the time of the inquiry, began to grapple with them and develop policy. Similarly, many genetic research laboratories were admirably proactive in reviewing their processes for securing consent, maintaining privacy, and so on.

Finally, the community debate and discussion that surround most law reform processes can contribute to the changing social atmospherics or *Zeitgeist* – or, as Oliver Wendell Holmes (1880) put it, "the felt necessities of the time" – that subsequently make possible parliamentary development of the statute law or judicial development of the common law. For example, the ALRC's report on the recognition of Aboriginal customary law was never formally implemented but might nevertheless have influenced the thinking of the majority of the High Court in *Mabo* (1992),[6] when it declared the existence of a form of customary Native title under Australian common law, after more than 200 years of resistance to this concept by the legal system.

The "Extra Step" of Widespread Public Engagement

A deep commitment to undertaking extensive community consultation as an essential part of research and policy development is the defining characteristic of a modern law reform commission. Ultimately, it is that attribute which distinguishes it from other bodies (think tanks, research institutes) with a law reform aspect to their work. Brian Opeskin (2002, 54) has noted that "the desirability of engaging the public in the process of law reform may be explained in many ways. For convenience, these can be divided into three groups: benefits for those consulted, benefits for the process of law reform, and benefits in terms of enhanced effectiveness of the law once reformed." From its earliest days, under the leadership of the foundation chair, Justice Michael Kirby, the ALRC entrenched the active engagement of the community as part and parcel of its basic approach to law reform, under the rallying cry that law reform is much too important to be left to the experts. The ALRC acknowledged that consultation had become an attribute of organized law reform since the creation of the English Law Commission in 1965; however, the ALRC committed itself to taking the extra step of ensuring, as a matter of basic philosophy, that law reform should be conducted in a transparent way with opportunities for widespread public consultation.

As Kirby (1986, 2-3) later reflected,

> For me, it was never a purely theoretical or analytical challenge. Law affected intimately the lives of people. To reform it, and thus to make it better, it was essential to consult the "usual suspects" – judges, legal practitioners, public officials and institutions. But it was also important to consult ordinary people. They might offer perspectives that would refine and strengthen our proposals. Moreover, the very process of consultation would build a momentum that would protect the ALRC against the risks of bureaucratic and political indifference when its reports were finally written and tabled in the Parliament.
>
> The process of widespread consultation was a reminder to the expert participants in the ALRC of the need to step beyond an elitist and purely lawyerly approach to law reform. Sometimes it added perspectives that the experts had missed, or identified sensitivities that needed to be addressed. Occasionally it repaired the imbalances between the well-organised lobby groups and the interests of ordinary people. It provided a forum to test expert ideas in civil society and to question intelligent laymen about their views and experience. Above all, it was a new scene: judges, lawyers and professors asking those affected about the law and how it could be made better.

Kirby noted even later that

> The ALRC technique symbolised its commitment to a non-elitist approach to law reform. It gave the agency a high public profile that helped to protect it from abolition. It raised expectations in the community of action in the area of law concerned. It made it more difficult for the government and the

Parliament to place the recommendations in the too hard basket. (Kirby 2005, 435-36)

Tailoring Consultation Strategies to Particular Issues

The nature and extent of community engagement are normally determined by the subject matter of the law reform inquiry. Areas that are, or are seen to be, narrow and technical tend to be of interest mainly to expert legal practitioners, industry associations, and government agencies. For example, ALRC inquiries into evidence law, marine insurance, and the architecture of federal jurisdiction fall into this category. However, other ALRC inquiries – such as those relating to children and the law, Aboriginal customary law, multiculturalism and the law, gender equality before the law, and the protection of human genetic information – involved high levels of interest and participation from the general public.

As mentioned above, the real and perceived independence of a law reform commission is crucial in providing the level of confidence needed for successful community consultation. At least as important, people also must believe that the time and effort involved in their participation in the law reform process are worthwhile – that is, they will be given a meaningful opportunity to be heard, and there is some reasonable prospect of achieving positive change.

The "traditional" approach of law reform commissions and other inquiries to consultation has generally followed a predictable – and, in earlier times, relatively successful – pattern. The active elements in the process involved

- producing community consultation papers, providing a directed opportunity for stakeholders and interested members of the general community to respond to particular questions and proposals (normally via written submissions), and thus to contribute to the ongoing discussion and influence policy development;
- conducting formal hearings in which interested parties can provide evidence; and
- organizing more informal meetings, workshops, and roundtables with experts (public and private sector), peak associations, community groups, and other stakeholders.

The more passive but nevertheless important element in the consultation process involved soliciting and receiving written submissions from stakeholders.

However, it is now generally recognized that law reform and policy development are increasingly crowded fields (Weisbrot 2005), with a constant stream of parliamentary inquiries, departmental and interdepartmental inquiries, task forces, royal commissions, green papers, and the like, so that "submission fatigue" is real. It is no longer sufficient for a law reform commission to publish an issues paper or discussion paper, perhaps schedule a few public hearings, and then sit back and wait for the raft of comprehensive, thoughtful, and beautifully crafted submissions from stakeholders to flow in.

Particularly in inquiries in which public participation is less naturally forthcoming, greater creativity is required in fashioning a consultation program, which might include extensive use of the mass media (including talk-back radio); surveys of particular memberships or participants; telephone hotlines; public opinion polls; and market research techniques, including focus groups.

Law reformers also need to be more creative and dedicated about reaching out to non-mainstream communities – such as Indigenous, non-English-speaking, and rural and remote communities – that traditionally have lacked the resources to engage fully with consultation and policy-making processes. For example, it is rare to find consultation documents produced in a variety of community languages (or even in plain English), though some law reform bodies have been more attentive to this issue than others.[7]

In 2009, as part of its Reconciliation Action Plan, the ALRC devoted considerable time and resources to the establishment of a standing Indigenous Advisory Committee,[8] comprised of prominent Indigenous leaders, lawyers, and social scientists from around the country, to advise commissioners and staff about the development of strategies, protocols, and procedures for more effective consultation with Indigenous communities and organizations; potential subject matter for future ALRC inquiries that would be of special significance to Indigenous communities; and increased opportunities for recruiting more Indigenous staff (legal and administrative), internships and training programs for Indigenous people, and opportunities for "twinning" and other support arrangements with Indigenous organizations (e.g., Aboriginal Legal Services, research institutes, and educational bodies).

Although it is still early in the law reform context, one might expect that law reform bodies will increasingly develop more creative and intensive ways to harness the potential of the Internet and other forms of electronic communication, including social networking sites – for example, by providing more interactive websites, including online questionnaires, discussion groups, and chatrooms. Public forums and consultations could be streamed to a much larger audience (domestically and internationally). Planned carefully and managed sensibly, such efforts can promote community debate and education, elicit views and information to assist in fashioning policy recommendations, and provide the law reform commission with a higher public profile, which in turn encourages further interaction down the track.

Virtually all law reform agencies in the industrialized world have informative and well-maintained websites, and some already go beyond this. For example, the ALRC now uses a Twitter feed to alert interested members of the community about forthcoming meetings and publications. Over the past few years, the ALRC has established a "Talk to Us" blog site for each of its recent inquiries, allowing members of the community to share experiences and provide comments and suggestions about reform and thus to engage with the ALRC – and with each other – in exploring the issues under consideration. Given the sensitivity of some of these areas, the discussion groups are fully moderated by ALRC staff. Similarly, the British Columbia Law Institute developed a Facebook presence in 2010, both to provide information to individuals interested in its work and to serve as a point of contact for the sixty or so institutional law reform bodies around the world.

Case Study: The ALRC Inquiry into the Protection of Human Genetic Information

In February 2001, the Australian government initiated a major inquiry into the protection of human genetic information that was led by the ALRC in association with the Australian Health Ethics Committee (AHEC) of the National Health and Medical Research Council (NHMRC). The Terms of Reference directed the ALRC and AHEC to consider, with respect to human genetic information – and the tissue samples from which such information can be derived – how best to protect privacy, guard against unfair discrimination, and ensure the maintenance of high ethical standards.

Given the array of contexts in which human genetic information can be collected and used, these three central concerns were explored across (*inter alia*) the oversight of scientific and medical human genetic research; public health planning and administration; practical delivery of clinical genetic services; law enforcement uses of DNA for forensic purposes; insurance underwriting; employment; sport; immigration; parentage and kinship testing; use of DNA for the construction of ethnic, racial, and Aboriginal identity (which the ALRC opposed); use of DNA evidence in criminal and civil courts; and management of genetic registers, tissue banks, and human genetic research databases (HGRDs).

The major challenge for the ALRC and other policy makers in this area is to find a sensible path meeting twin goals: to foster innovations in genetic research and practice that serve humanitarian ends, and to provide sufficient reassurance to the community that such innovations will be subject to proper ethical scrutiny and legal (and other) controls. Put another way, the ALRC saw its task as developing a policy platform that would guard against wilfully or negligently bad practice – and thus avoid a breach of public trust of such magnitude that it would occasion a backlash against even ethically sound and scientifically valuable work. Successfully completing this mission involved not only providing adequate protections against the *unlawful* use of genetic information but also putting into place measures and strategies aimed at ensuring a higher-order goal: where such information might be used lawfully, it will be used properly, fairly, and intelligently.

The project culminated in May 2003 with the launch in the Australian Parliament of the report *Essentially Yours: The Protection of Human Genetic Information in Australia* (ALRC and AHEC 2003), which made 144 recommendations for reform. In keeping with the ALRC's strategic approach to these complex issues, the recommendations were addressed to the Australian government and to thirty other actors – including, among others, health and medical policy makers (e.g., the NHMRC, university medical school deans, and state and territorial health officials); doctors and allied health professionals; anti-discrimination and privacy officials; regulatory authorities (for consumer protection and regulation of medical devices); insurers; employers; and educators.

On 9 December 2005, a whole-of-government response to *Essentially Yours* was released; it accepted the overwhelming majority of the findings and recommendations (NHMRC 2003).[9]

The report has also gained significant attention and praise internationally. For example, in his keynote address opening the XIXth International Congress on Genetics in Melbourne in July 2003, Dr. Francis Collins – then director of the (US) National Human Genome Research Institute and head of the Human Genome Project consortium – described *Essentially Yours* as "a truly phenomenal job, placing Australia ahead of what the rest of the world is doing" (cited in ALRC 2003).

From the beginning, both the government and the ALRC recognized the critical need for public engagement and widespread consultation that would draw in the general community as well as experts and interest groups. For its part, the government specified in the Terms of Reference, Part 2(b), that the ALRC and AHEC should "identify and consult with relevant stakeholders, including the Privacy Commissioner and the Human Rights and Equal Opportunity Commission, and ensure widespread public consultation." Although exhortations to widespread community consultation are boilerplate language in most Terms of Reference for law reform inquiries, in this instance, the Australian government took the unusual step of backing that up with an additional grant of $400,000, on top of the ALRC's recurrent budget, to facilitate community engagement.

Apart from the normal production of a series of community consultation documents, the ALRC also designed a series of shorter brochures and pamphlets aimed at a general audience – and even experimented (with limited success) with distributing an eye-catching free postcard at restaurants and cafés, directing interested people to the ALRC's website and suggesting ways in which individuals could have their say on the protection of human genetic information.

The ALRC also conducted 15 major public forums in the capital cities and major regional centres, arranged about 250 "targeted" meetings with key stakeholders and community organizations in Australia and overseas, and received in excess of 350 substantial written submissions. The public forums were well publicized by paid advertisements in newspapers and on radio but especially by the extensive newspaper, radio, and television coverage of the ALRC's activities, and the forums were well attended. Each forum featured a brief talk from a leading geneticist or bioethicist, and from the ALRC president or commissioner, outlining the medical and scientific background as well as the ethical, legal, and social implications (ELSI) confronting the inquiry, and then ample time was provided for questions and comments from the audience. The forums were scheduled to run for about two hours each, but, given the high level of interest and participation, most ran much longer.

Not surprisingly, the "ALRC road show" received greater coverage in regional areas, with less competition for media attention, than in the major cities – the single biggest crowd was in Townsville, in far north Queensland, dwarfing the attendance in Brisbane, the state capital and largest city. In Hobart, Tasmania, many members of the audience indicated that they had driven for hours from outlying areas to attend the evening forum. The ALRC also intentionally varied the venues in an effort to attract different audiences. For example, in both Melbourne and Perth, the forums were held in the auditoriums of major teaching hospitals and attracted a large number of researchers and health professionals. In other cities, the ALRC rented function rooms in downtown hotels and attracted a higher proportion of members of the general public, some of them with personal or family experience of genetic conditions, and some people who had heard the advance publicity and were merely curious to learn more about the emerging science and related ELSI issues. The ALRC's meetings in Darwin and Alice Springs, in the Northern Territory, were the most sparsely attended but contained by far the highest representation of Indigenous people, and the meetings were valuable for that reason.

Dealing with Stakeholders

During the genetic information inquiry, the ALRC sought opportunities to bring together (with mixed success) population geneticists and health researchers with privacy advocates; business and employer groups with trade union officials; custodial mothers with "fathers' groups" in relation to DNA paternity testing; and law enforcement authorities and prosecutors with defence lawyers and civil libertarians.

However, it would be misleading to imply that simply facilitating opportunities for civil discussion among competing stakeholders is always – or even often – likely to result in a happy consensus about policy development on contentious issues. The head of a high-profile inquiry must carefully manage stakeholder expectations, including which subject areas might be considered within the scope of the Terms of Reference.

The community consultation aspect of public policy formation is not a simple numbers game, so that a well-written personal or organizational submission is likely to have much more impact on the ultimate recommendations and advice to government than a letter-writing or post card campaign in which large numbers of individuals parrot the same few phrases. There are, of course, some opportunities for a popular vote on policy initiatives – for example, at a general election or especially in relation to a referendum – but community consultation on law reform and social policy is not such an exercise.

It is equally important to move the community debate away from the polarizing and unhelpful language of (absolute) rights. In fact, as the ALRC has expressly acknowledged in a number of its recent reports, "privacy interests unavoidably will compete, collide and coexist with other interests. For example, privacy often competes with freedom of expression, a child's right to protection from abuse, national security considerations and so on. No single interest – not even one elevated to the status of a human right – is absolute" (ALRC 2008, para. 1.57).

Handled properly, however, a further benefit of open and extensive public consultation is that the process clearly identifies competing arguments, interests, and groups – and their relative support in the general community – well in advance of any governmental action. Public ventilation of issues, including the opportunity for education, debate, and participation, is sometimes enough to defuse lingering tensions.

Sometimes a broad consensus can be achieved through the law reform process; more often, though, a commission has to make hard decisions and identify a preferred approach. A law reform commission's final report is not an advocate's brief; rather, it should set out for the government's consideration a thorough analysis of the relevant law and policy; identify the various models and options considered; document the range and relative strength of the opinions encountered (through meetings, submissions, and a review of the literature); and then provide sound reasons for its recommendations.

Whether or not a government ultimately agrees with the approach taken by the commission, an open inquiry should guarantee that the government is never taken by surprise when it engages in law making in the area. Opposition criticism might also be more muted where an initiative has the imprimatur of an independent law reform commission and is thus seen to be non-partisan.

Media Management

Although media coverage is often critical to stimulating public interest in and engagement with law reform consultation processes, commissions need to develop media management skills and strategies to ensure that they do not lose control over the presentation of issues. It is unfortunate for proper public debate and policy making, but media coverage typically prefers personal conflict to the contest of ideas; simple explanations to complex ones; and striking images to ideas, however interesting.

There is also strong media resistance to the iterative and deliberative approach favoured by law reform commissions. Although a final report with recommendations for change fits the media paradigm – reporters can readily find experts and representatives of interest groups who will argue passionately for or against the particular recommendations – it is more difficult to gain effective media coverage of the preceding community consultation papers, where the objective is to raise consciousness and develop issues and ideas for further consideration.

Even with final reports, it is sometimes difficult for a law reform commission to convey more than a single message – and that message might not always be the one that the commission itself rates most highly. For example, media coverage of the landmark *Essentially Yours* report was disproportionately skewed toward the emotive issue of DNA paternity testing – a hot-button issue for fathers' groups – covered in one of the report's forty-six chapters (ALRC and AHEC 2003, Chapter 35). The media thus missed or glossed over most of the major issues and recommendations, including some that had been covered in the lead-up to the final report, such as genetic discrimination in the workplace.

In retrospect, it might have been more effective, from a media strategy point of view, to release a series of final reports over time to secure better coverage of the many important issues. However, this would have meant fragmenting some of the research and general discussion, such as the important consideration of underlying themes and overarching principles. This would also have meant withholding from government for a period of time some of the findings and recommendations, after they had been finalized, simply for media purposes – which the ALRC was not prepared to do.

Conclusion

Effective community consultation is expensive, time consuming, repetitive, and exhausting. However, based on my long experience in institutional law reform, I have no doubt that consultation that is well crafted and carefully tailored to the particular issues and stakeholder communities ultimately results in much better policy formulation. This is especially the case where the reform effort is intended to have an impact on the practical, grassroots level – where almost all individual interactions with government and industry occur – rather than to produce high-order advice or statements of principle.

The stories and case studies elicited through community consultation are valuable in identifying specific issues and concerns that will not likely be raised at the peak association or government officer level. Interactions with the community also help to identify the "common sense" of the community to find common ground among different stakeholder groups. Effective consultation means that stakeholders will think that they have genuinely been "listened to" – critical for subsequent legitimacy, whatever the final call in terms of policy advice.

For example, the consultation effort associated with the ALRC's inquiry into the protection of human genetic information involved thousands of hours of discussions with affected families, genetic support groups, genetic counsellors, familial cancer registries, family doctors, and others, which led to detailed recommendations targeting resources in this area.

It is one thing to make a bland recommendation that, for example, "individuals should not be unfairly discriminated against by reason of their genetic status." It takes a great deal more digging to determine

- the degree to which some public resources already exist for these purposes;
- the adequacy of these resources and the delivery of existing programs, including in more remote or marginalized communities;
- whether there is access to and equity in genetic services;
- whether there are roadblocks or pinch-points in the delivery of genetic services and counselling, such as through inadequate coordination among different government departments or federal, state, and local authorities;
- whether the "gatekeepers" of the health-care system – family doctors and other front-line health-care officers – are sufficiently educated about genetic medicine to provide meaningful assistance to their patients;
- whether there are cultural issues that inhibit access to genetic services and counselling (e.g., unwarranted feelings of shame, or the unwillingness of women in some communities to speak candidly with male health professionals);
- the nature and extent of primary and continuing medical education about genetics and which institutions (university medical schools? medical professional associations? royal colleges?) should play a role;
- whether genetic counsellors require particular recognition by the health-care system (whether by statute or by administrative mechanisms) to allow their fees for services to be funded or rebated through the universal Medicare system;
- the extent to which community organizations, such as genetic support groups, should receive public funding and support, given their positive role as reported by affected families; and
- the best method of organizing all of these services, so that the health-care system can take strategic advantage of efficiencies of scale and assembling of expertise, while patients can be assured of access to high-quality and cost-effective services.

This level of policy development and reform is only possible through extensive, carefully designed consultation with members of the community with daily experience of these matters, whether as service seekers or service providers. A "tick-a-box" consulting exercise, done only to satisfy a government pledge to "consult," can never deliver detailed and practical recommendations reflecting community experiences and priorities, nor can it deliver a sense in the community that it has been respectfully and meaningfully included in the policy development process.

NOTES

1 The legislation was specified to take effect on 1 January 1975, allowing a period for recruitment and the logistics of creating a new institution.
2 As exemplified by the Law and Development and Law and Modernization movements in the United States and United Kingdom in the 1960s and 1970s.
3 Papua New Guinea is somewhat unusual insofar as its law reform commission is accorded constitutional status in its 1975 Independence Constitution, section 21 and Schedule 2.13-2.14.
4 *State Government Insurance Commission v. Trigwell* (1979), 142 CLR 617, c. 22.
5 A few law reform agencies operate successfully within government. For example, given their small size, a few Pacific Island states have vested the law reform function in an existing senior officer, such as the solicitor general (as in Tonga) or the attorney general (as in Samoa, before an independent law reform commission was created by statute in 2002 and funded and staffed several years after that). In Singapore, the Law Reform and Review Division of the Attorney-General's Chambers operates as the nation's main law reform agency; although lacking statutory or structural independence, it has an excellent reputation for the quality of its work.
6 *Mabo v. Queensland* (No. 2) (1992), 175 CLR 1.
7 A positive example is the Victorian Law Reform Commission in Australia, which provides summaries (at least) of major consultation documents in a range of community languages and offers this facility on its website: http://www.lawreform.vic.gov.au.
8 For more information, see http://www.alrc.gov.au/.
9 Roughly 90 percent of the ALRC's recommendations were accepted by the government in whole or in substantial part.

REFERENCES

Australian Law Reform Commission (ALRC). 1996. *Australian Law Reform Commission Act 1996 (Cth) ss 21 and 24.* http://www.alrc.gov.au/.
–. 2003, 14 July. *ALRC Media Release: ALRC Work Praised at World Genetics Congress.* http://www.alrc.gov.au/.
–. 2008. *ALRC Report 108: For Your Information – Review of Australian Privacy Law.* http://www.alrc.gov.au/.
Australian Law Reform Commission (ALRC) and Australian Health Ethics Committee (AHEC). 2003. *ALRC 96 Essentially Yours: The Protection of Human Genetic Information in Australia.* http://www.austlii.edu.au/.
British Columbia Law Institute. 2008. Law Reform Database. http://www.bcli.org/.
Burg, E. 1977. "Law and Development: A Review of the Literature and a Critique of 'Scholars in Self-Estrangement.'" *American Journal of Comparative Law* 25: 492-530.

Handford, P. 1999. "The Changing Face of Law Reform." *Australian Law Journal* 73: 503-23.

Holmes, O.W. 1880. *The Common Law.* http://www.1215.org/.

Kirby, M. 1986. *The Speeches of the Honourable Justice M.D. Kirby, CMG,* Vol. 1, *1975-1976.* Sydney: Australian Law Reform Commission.

–. 2005. "Are We There Yet?" In *The Promise of Law Reform,* edited by B. Opeskin and D. Weisbrot, 433-36. Sydney: Federation Press.

Mason, A. 1975. "Where Now?" *Australian Law Journal* 49: 573-75.

Merryman, J. 1977. "Comparative Law and Social Change: On the Origins, Style, Decline, and Revival of the Law and Development Movement." *American Journal of Comparative Law* 25, 3: 457-91.

National Health and Medical Research Council (NHMRC). 2003. *Australian Government Response to* Essentially Yours: *The Protection of Human Genetic Information in Australia.* http://www.nhmrc.gov.au/.

Opeskin, B. 2002. "Engaging the Public: Community Participation in the Genetic Information Inquiry." *Reform* 80: 53-58.

Sawer, G. 1970. "The Legal Theory of Law Reform." *University of Toronto Law Journal* 20, 2: 183-95.

Tilbury, M. 2005. "A History of Law Reform in Australia." In *The Promise of Law Reform,* edited by B. Opeskin and D. Weisbrot, 3-17. Sydney: Federation Press.

Trubek, D., and M. Galanter. 1974. "Scholars in Self-Estrangement: Some Reflections on the Crisis in Law and Development Studies in the United States." *Wisconsin Law Review* 4: 1062-1102.

Weisbrot, D. 1990. *Australian Lawyers.* Melbourne: Longman Cheshire.

–. 2005. "The Future of Institutional Law Reform." In *The Promise of Law Reform,* edited by B. Opeskin and D. Weisbrot, 18-22. Sydney: Federation Press.

11

Participation in the Canadian Biotechnology Regulatory Regime

ANDREA RICCARDO MIGONE AND
MICHAEL HOWLETT

The Relevance of Participation

The increasing relevance of public participation in industrialized countries (Brunk 2006; de Jonge et al. 2008; Peters et al. 2007; Wynne 2006) can be framed from the vantage point of institutional contexts to assess the nature and extent of this participation and to individuate how institutions foster, direct, or limit participation itself. It is of particular interest to focus on policy and regulatory regimes to address these questions.

A sector where public participation can be particularly important is biotechnology because applications in both the medical and the non-medical fields can generate complex ethical, health, and economic issues. Policy makers face a variety of potential issues when they engage biotechnology policy regimes. These issues include protecting personal genetic information, establishing and enforcing appropriate health and environmental protection standards, and designing tools to balance market development and consumer protection and information alike. The notion that biotechnology is a critical area for public engagement is borne out by the increasing use of Danish-style consensus conferences in countries such as Norway, the Netherlands, France, Japan, South Korea, New Zealand, the United Kingdom, and the United States (Seifert 2006, 77).

The reception of biotechnology in general (Coyle and Fairweather 2005; Hornig Priest 2006; World Health Organization 2005), in the medical field (Avard, Grégoire, and Jean 2008; Greely 2001), and of genetically modified foods in particular (Andrée 2006; Durant and Legge 2006) highlight these concerns and the need for an analytical/educational approach that minimizes the negative impacts that these technologies have on public perceptions of products under development. An important factor in the acceptance of these technologies is the level of effective participation that the public and stakeholders have been allowed in the process (Avard, Grégoire, and Jean 2008). For example, Gutteling and colleagues (2006, 111) found that in the Netherlands, "trust is related to the way government or politicians are inclined to involve the public within decision-making, how the industry is handling consumer interests, and individuals' perception of the way biotechnology may influence their life." Another case in which public debate assumed an important dimension in the area of genetically modified foods was that during 2002 in Zambia regarding genetically modified food aid (Mwale 2006). A recent survey notes how an important section of the US and Canadian population wants to have a voice in the debate on gene technology (Hornig Priest 2006). Beyond these points, what Sharp, Yudell, and Wilson (2004, 3) call constructive catalysts, the process of focusing attention on a particular event to influence policy change, can be very important for biotechnology. A crucial part would be played in this case by popularizing the research and opening up discussion to a broad range of stakeholders.

Calls for a more participatory and informative approach to the diffusion of biotechnology have been made by various groups (Abelson et al. 2007; Avard, Grégoire, and Jean 2008; Haga and Willard 2006; Pew Initiative on Food and Biotechnology 2006; Sharp, Yudell, and Wilson 2004). This might be relevant to more than just inclusive policy making since the attitudes of consumers are related to what they know about the benefits and risks associated with GM foods (Brown and Qin 2005) and to their level of information (Costa-Font and Mossialos 2005). In the United States, the public has been shown to be relatively segmented on the issue of GM food, and the effect of labelling on such foods is showing different results (Teisl, Radas, and Roe 2008). Public opinion in Canada on such matters is nuanced in its understanding of public policy in the field of genetics, but there is a call for weighing the benefits and drawbacks of these technologies (Hornig Priest 2006). A call for more deliberative dialogue in the country, in place of the usual "polling and adversarial dialogue," was put forward by the Canadian Biotechnology Advisory Committee (BSDE Expert Working Party 2006, 30).

We believe that in Canada, the process of participation related to the field of biotechnology is relatively advanced in its implementation, for most governmental agencies have engaged in public consultations, and the policy processes are relatively transparent and open. Here we use the differentiation noted by Castle and Culver (2006) between engagement (a process largely limited to informing the public) and consultation (a process that actually considers the public's opinion). However, we argue that in Canada participation of either kind has to date had relatively little effect on the type of policy regime that has emerged, being mostly limited to what appears akin to a voice option (Hirschman 1970).

The Canadian Biotechnology Policy Regime

As Kleinman et al. (2009) noted, policy regime strategies and regulatory frameworks in the biotechnology sphere are interconnected, the latter delivering the needed detail in enforcing and fostering a specific direction and set of goals for national science and technology policies. These policy regimes emerge at the national level, though they are closely connected with international developments in terms of their links with regulatory frameworks such as the Cartagena Protocol on Biosafety (Newell 2008) or the Codex Alimentarius in the area of food safety (Lindner 2008). Within this common international policy space, however, different countries regulate, foster, and support biotechnology in different ways (Lindner 2008).

On this basis, Isaac (2001) for example, has argued that North American and European Union approaches to the regulation of agricultural biotechnology differ according to their diverging interpretations of the precautionary principle. North American jurisdictions highlight and prioritize scientific concerns, and European countries tend to emphasize social concerns and responsibilities (see Table 11.1).

This analysis is general and requires a more fine-grained approach if local variations in biotechnology regulation are to be understood. This is especially true in the context of a shifting pattern of regulatory behaviour brought about by the extension of biotechnology activity away from an emphasis on GMOs and toward a less interventionist but much broader application of genomics, metabolomics, transcriptomics, and proteomics. Genomics as the study of an organism's genome is not to be confused with genetic manipulation. Although relatively little government or political hindrance is posed to the study of the genome's structure, the application of this knowledge to genetic manipulation is much more politically difficult. The two areas can be considered cognate since genomics research is one of the prerequisites of genetic manipulation. There is little regulation directed specifically toward the former, whereas the latter is both regulated and often perceived as dangerous by the public. There is little doubt, however, that

TABLE 11.1

Scientific versus social rationality in genomics regulation

	Scientific rationality (North America)	Social rationality (Europe)
General regulatory issues		
Belief	Technological progress	Technological precaution
Type of risk	Recognized Hypothetical	Recognized Hypothetical and speculative
Substantial equivalence	Accepts SE	Rejects SE
Science or other factors in risk assessment	Safety Health	Safety Health Quality Socio-economic factors
Burden of proof	Innocent until proven guilty	Guilty until proven innocent
Risk tolerance	Minimum risk	Zero risk
Science or other factors in risk management	Safety or hazard based: risk management is for risk reduction and prevention only	Broader socio-economic concerns: risk management is for social responsiveness
Specific regulatory issues		
Precautionary principle	Scientific interpretation	Social interpretation
Focus	Product based, novel applications	Process or technology based
Structure	Vertical, existing structures	Horizontal, new structures
Participation	Narrow: technical experts Judicial decision making	Wide: "social dimensions" Consensual decision making
Mandatory labelling strategy	Safety or hazard based	Consumers' "right to know" based

Source: Isaac (2001, 2).

both are beginning to attract increasing interest from the public and that this interest is not always benign or unconcerned.

Haga and Willard (2006) provide some of the details required to understand and explore the regulatory activity undertaken in this emerging area of public policy. They argue that we can identify five types of regulatory issue fields in the biotechnology sector, such as health, commercialization, and intellectual property rights, which intersect the issue areas. How these legal issues, the public research investment, and the ways in which risk management and regulatory oversight in the policy deliberation process are handled determines the key features of these regimes (Talukder and Kuzma 2008, 131).[1] Table 11.2 provides a list of eight basic issues that they identify with which biotechnology regulation has grappled, along with the five dimensions that each involves.

Regulatory policy making in the field of biotechnology involves the design and adoption of a set of policies that will deal with the issues noted in Table 11.2 and will dovetail with country-specific circumstances in the sector. Viewed in this light, there are substantial differences among countries that are lost in Isaac's model. Two good examples of such variations are in the variance between US and Canadian GMO policies (Montpetit 2005) and between agricultural and medical GMOs within both countries (Sheingate 2006).

Paarlberg (2000) used a similar system to score issue areas linked to first generation biotechnology policy to generate a country (or sector) measure that ranged policy approaches in terms of their "promotional," "permissive," "precautionary," and "preventive" nature. Policies that accelerate the spread of GM crops and food technologies within the borders of a nation can be termed "promotional." Policies that are neutral toward the new technology, intending neither to speed nor to slow its spread, are called "permissive." Policies intended to slow the spread of GM crops and foods for various reasons are termed "precautionary." Finally, policies that tend to block or ban entirely the spread of this new technology are defined as "preventive" (Paarlberg 2000, 4) (see Table 11.3).

Haga and Willard's work (2006) play an important part here on another level, for they highlight the importance of risk management and regulatory oversight in the policy deliberation process (Talukder and Kuzma 2008, 131), a topic that Paarlberg (2000) did not consider. Participation by the public and stakeholder groups in the policy process when new technologies are involved is a subject that has received a great deal of attention in recent years (Haga and Willard 2006; Sharp, Yudell, and Wilson 2004; Tutton

TABLE 11.2

Regulatory issue field in biotechnology

Issue areas	Research issues	Legal issues	Economic issues	Education issues	Acceptance and implementation issues
Intellectual property rights	Patent policy	Intellectual property and licensing practices	Cost effectiveness		Acceptance of biotech private ownership
Public information/ inclusiveness of deliberation	Ethics review	Privacy and confidentiality	Cost of broad consultations Intellectual property	Development of clinical guidelines Classroom education Public education Risk communication	Behavioural modification in response to biotechnology results
Commercialization and retail trade	Patent law	Trade agreements	Market value and pricing Supply and demand Commercialization of public-sector initiatives Creation of new market segments	Labelling	Public adoption of biotechnology
Food and health safety	Creation of a regulatory framework	Regulatory oversight (product and manufacturing review, labelling, laboratory quality, and environmental impact)	Costs related to testing	Education of health professionals	Acceptance of the safety of food products by the public

Human health	Creation of a regulatory framework	Regulatory oversight (product and manufacturing review, labelling, laboratory quality, and environmental impact) Issues of privacy Genetic discrimination	Market value and pricing versus public provision of health care Costs related to testing	Education of health professionals	Acceptance of the safety of health products by the public
Consumer choice	Media advertising	Genetic discrimination	Different responses in consumer behaviour	Information directed to consumers	Cultural respect
Public research investment	Prioritization of research areas (basic, applied, and technology development) Allocation of funds Provision of facilities Access to tools and research samples	Protection of human subjects Ownership of research results	Research and development funding Economic incentives for biotechnology research	Information directed to citizens	Acceptance of the value of biotechnology investment
Commercialization of biotechnology-related products	Reliance on university-generated research Patent policy	Intellectual property rights	Accessing venture capital Creation of technology licensing organizations	Labelling Pedagogical research	Acceptance of the value and safety of biotechnology products Public opinion research

Source: Haga and Willard (2006, 967).

TABLE 11.3

Paarlberg model of policy options and regimes toward GM crops

	Promotional	Permissive	Precautionary	Preventive
Intellectual property rights	Full patent protection, plus plant breeders' rights (PBRs) under UPOV 1991	PBRs under UPOV 1991	PBRs under UPOV 1978, which preserves farmers' privilege	No IPRs for plants or animals or IPRs on paper that are not enforced
Biosafety	No careful screening, only token screening or approval based on approvals in other countries	Case-by-case screening primarily for demonstrated risk, depending on intended use of product	Case-by-case screening also for scientific uncertainties owing to novelty of GM process	No careful case-by-case screening; risk assumed to be cause of GM process
Trade	GM crops promoted to lower commodity production costs and boost exports; no restrictions on imports of GM seeds or plant materials	GM crops neither promoted nor prevented; imports of GM commodities limited in same way as non-GM in accordance with science-based WTO standards	Imports of GM seeds and materials screened or restrained separately and more tightly than non-GM; labelling requirements imposed on import of GM foods or commodities	GM seed and plant imports blocked; GM-free status maintained in the hope of capturing export market premiums
Food safety and consumer choice	No regulatory distinction drawn between GM and non-GM foods when either testing or labelling for food safety	Distinction made between GM and non-GM foods on some existing food labels but not so as to require segregation of market channels	Comprehensive positive labelling of all GM foods required and enforced with segregated market channels	GM food sales banned or warning labels that stigmatize GM foods as unsafe to consumers required
Public research investment	Treasury resources spent on both development and local adaptations of GM crop technologies	Treasury resources spent on local adaptations of GM crop technologies but not on development of new transgenes	No significant treasury resources spent on either GM crop research or adaptation; donors allowed to finance local adaptations of GM crops	Neither treasury nor donor funds spent on any adaptation or development of GM crop technology

Source: Paarlberg (2000, 4).

2007). Considering that public perceptions of and attitudes toward genomics/GMOs are often confused (Fischhoff and Fischhoff 2001), including this dimension in the analysis is beneficial.

Specifically in terms of regulatory tools, Haga and Willard (2006) argue that the policy issue areas for GMO technology have been tackled with one, or a mix, of the regulatory approaches listed in Table 11.2 (see Table 11.4). These regulatory approaches can be synthesized within two broad categories (state versus public approaches) and linked to Paarlberg's categories to create the comparative national matrix shown in Figure 11.1. State approaches are based on scientific rationality and include legislative, regulatory, and guideline approaches. Public approaches are based on voluntarism or public consultation. Starting from this matrix, we can rank countries or sectors within four policy quadrants reflecting the preferences shown in utilizing either elite or public policy approaches in their dealings within the subject area. An example of the application of this tool for Canada is given in Table 11.5.

TABLE 11.4

Haga and Willard list of regulatory approaches to GM issues

Approaches	GM issues
Legislative	*Genetic discrimination:* more than twenty bills introduced in the United States to prohibit genetic discrimination by health insurers and/or employers
Regulatory	*Genetic testing:* proposal to revise the US Clinical Laboratory Improvement Amendments regulations to add the quality of genetic testing as a specialty
Guidelines	*Gene patenting:* revisions to the utility criteria of the US patent examination guidelines
	Licensing: US National Institutes of Health have published best practices for the licensing of genomic inventions
Voluntary	*Genetic discrimination:* Association of British Insurers Concordat and Moratorium on Genetics and Insurance
	Genetic testing: establishment of EuroGenTest Network to ensure quality of tests
Public consultation	*Genetic discrimination:* an eighteen-month public consultation carried out by the Australian Law Reform Commission
	GM foods: GM Nation public dialogue in the United Kingdom

Source: Haga and Willard (2006, 968).

FIGURE 11.1

Comparative biotechnology regulatory regimes

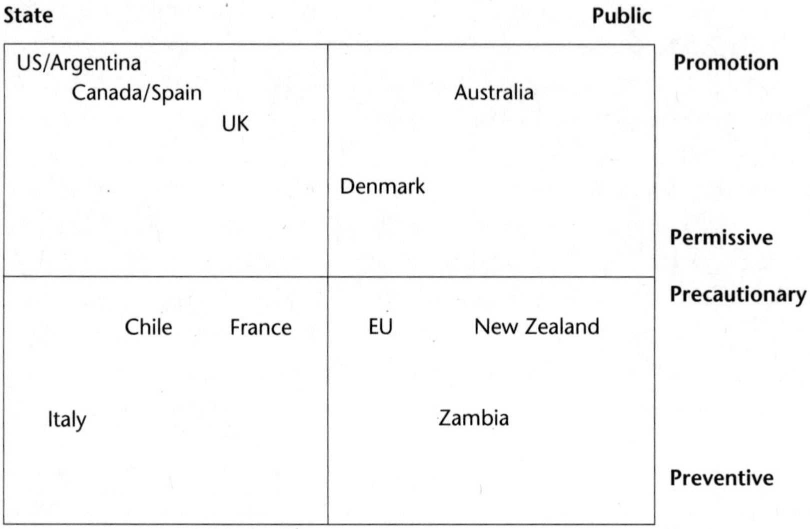

State				Public
US/Argentina Canada/Spain UK		Australia Denmark		Promotion
				Permissive
				Precautionary
Chile France		EU New Zealand Zambia		
Italy				Preventive

Denmark, for example, can be seen to marry a tradition of using consensus conferences through which public comments on biotechnology are formulated (Seifert 2006) with rather strict guidelines regarding cloning and a more open genomics research aspect. During 2001-2, Zambia instituted a process of public debate that led to the refusal to accept a shipment of what might have been GMO corn, thus falling into the publicly driven preventive quadrant (Mwale 2006). In the European Union, the leading principle in food safety policy has been the precautionary one (Lindner 2008, 142) mixed in with at least some examples of increased bottom-up models (Seifert 2006). This has partially changed since 2004, when the World Trade Organization found that the European Union had implemented a de facto moratorium on GMO products. Since then, various types of GMO corn and (in March 2009) the now obsolete T45 type of canola were approved for import into the European Union. In general, however, countries such as Greece, Italy, and France have maintained a negative attitude toward GMOs.

We argue that institutional settings influence the effects of participation. In the simplest policy cycle (agenda setting, policy formulation, adoption, implementation, and evaluation), participation is likely to have important

FIGURE 11.2

Distribution of participation instruments

State		Public
Polling	Commissions	Public consultation
Request for feedback		Consensus conferences

effects in agenda setting, policy formulation, and evaluation. Yet, it is important to distinguish between the instruments of participation and the effects of participation on the policy. We also argue that, while the instruments chosen to channel participation affect the eventual shape of a policy (say by choosing to limit participation to a request for general feedback or by expanding it through consensus conferences), the general rationale of the policy regime in which this participation occurs influences the type of participation instruments chosen. This echoes the effects of institutional settings on policy instruments. We believe that, in a biotechnology policy regime leaning toward social rationality, we might see more instruments aimed at participation, whereas a scientific rationality model is likely more preoccupied with educational efforts and would be more likely to show limited inclusion in the consultation process.

We see participation tools as policy instruments, and, according to our model, we would expect the use of specific participation instruments in biotechnology to be correlated to the state/public dimension, with the public side being more likely to see real participation as opposed to engagement. If we consider a set of participation instruments ranging from polling, to requests for feedback, to the constitution of commissions or expert groups, to public consultations and consensus conferences, we would expect them to be arranged more or less as shown in Figure 11.2.

It is harder to place these instruments on the promotional to preventive dimensions because their ultimate use depends on the basis on which biotechnology is perceived in a certain country. For example, referenda have tended to promote increased state intervention in biotechnology regulation (Rothmayr and Varone 2009). It seems fair to assume, however, that regimes that rely heavily on a scientific rationality might be more inclined to use a more state-centred approach in the selection of participation instruments, whereas ones that focus on social rationality might be more comfortable with public ones.

We argue that the Canadian biotechnology sector can be summarized in terms of this analysis in the results contained in Table 11.5, offering a quasi-promotional environment for the development of biotechnologies and relying mostly on a guideline style for regulation.

In terms of biotechnology regulation, Canada positioned itself closer to the open approach chosen by the United States than to the less permissive one typical of the European Union (Cantley 2007). The early phase of biotechnology adoption and regulation (between the mid-1970s and the mid-1980s) saw important gains in the development and expansion of the technology and in the acceptance and commercialization of its products, progressively relaxing the relevant regulatory frameworks. Since 1994, Health Canada approved over 100 novel foods, many of them involving genetic manipulation (Canada 2007, 5).[2] The promotional approach to biotechnology is reflected in various areas.[3] For example, only in 2004 did the Canadian General Standards Board produce a voluntary labelling standard for genetically modified foods in which genetically engineered material is over 5 percent of the product.[4] Although generally far from EU standards, this approach is still stricter than the one in place in the United States. In the Supreme Court of Canada Harvard Mouse case decision,[5] the balance partially shifted toward the social rationality principle, leading to tighter regulations and more economic difficulties and conditions for firms engaged in the development and use of biotechnology. Also, the Canadian testing process and its triggers remain more restrictive than the American ones. Although formally applying a substantial equivalency risk assessment principle (Canada 2001, 11), the Canadian regulatory process is still tougher and more broadly geared toward checking the nature of new GM products than the US one. This reflects the hybrid nature of the overall approach to the production and commercialization of biotechnology, with promotional research and commercialization processes and a permissive testing side.

Participation as Voice: The Canadian Experience

What are the spaces reserved for, and the efficacy of, participation in the Canadian biotechnology system? Canada uses a product-based approach to the regulation of biotechnology. This means that the Canadian regulatory system is highly geared toward the detection and regulation of novel traits, and many new techniques, such as marker-assisted selection (MAS) and other genomics tools, therefore fall outside the scope of most regulation.

TABLE 11.5

The Canadian biotechnology sector policy regime

Level	Operating element	Implementation processes
Policy regime	Quasi-promotional approach with mainly top-down scientific risk assessment	Permissive with elements of precaution in testing and screening of novel foods Promotional in the public research, IPR, and consumer choice areas Promotional/permissive in the trade area
Regulation	Guidelines style within "novel traits" regulatory approach	Preference for incorporating legislative and regulatory tools for biotechnology in existing legislation and regulation Equating biotechnology products with non-biotechnology ones Voluntary labelling for GMOs Guidelines for the specialized agencies that supervise and foster biotechnology development
Innovation	Industrial complex to Italianate district model	Canada tried to foster the creation and market application of biotechnology in keeping with the original framing of the field as an economic opportunity. This attitude is visible in the goals of the federal science and technology policy. The practical implementation of this vision passed through important research funding and investment and research incentives for the private sector. Results have been mixed: for example, the choice of supporting multiple biotechnology research centres across Canada did not result in multiple successes.
Participation	Participation instruments correlated to state-centred approach	Efforts to educate the Canadian public have been mixed, with limited engagement and relatively little policy change that was not generated by the federal government (i.e., voluntary approaches to GMO disclosure).

The Plant Biosafety Office (PBO) decides whether a plant with novel traits can be released into the environment.

Because of the relevance for research and especially commercialization of public support of these technologies and their applications, more effective involvement of the public might be important. The Canadian Biotechnology Advisory Committee has noted that public confidence in the process through which new technologies are introduced is critical to their acceptance (BSDE Expert Working Party 2006, 16). The same report called for a more deliberative dialogue to be implemented in Canada (30). The federal government has tried to address these concerns with projects such as the Biotechnology Notices of Submission Project, within which the CFIA posts on its website the notices of submission for GMO products and allows for submissions from the public. The questions received are then explored by the CFIA or Health Canada if they are of a scientific nature or streamlined into a less specific area if they are not.[6] This project started in 2003 and is still active. Again in 2003, Health Canada asked for public input into the revision of its Guidelines for the Safety Assessment of Novel Foods and, in 2005, for an options analysis paper on the Environmental Assessment Regime for New Substances in Products Regulated under the Food and Drugs Act.[7]

However, counter to what Canadians would like to see happen (Longstaff, Burgess, and Lewis 2006), participation in Canada is largely limited to venues that do not engage the public in the key central stages of the policy process and in meaningful negotiation over the content of policies. Although it is important to notice that the process of consultation in the Canadian biotechnology sector is well developed, involving both simple engagement and consultation (Castle and Culver 2006), we argue that the system remains akin to a "voice" option rather than true consultation. For example, though Castle and Culver (2006) argue correctly that there was proper consultation in the process that led to the voluntary labelling of GMO foods in 2004, we see this as a minor change in the overall policy regime structure. Regarding this issue, consumer demand in Canada seemed to be ahead of the regulatory curve. In 1999, under pressure from the public, the Canadian Council of Grocery Distributors launched an initiative to create a national labelling standard (outside the Food and Drugs Act) to give more information to Canadians regarding the content of their food, which resulted in the 2004 voluntary standard. The question remains whether the latter is an efficient or even broadly legitimate tool given that many groups that supported mandatory labelling did not participate in the process. While waiting for the standard to emerge, organic labelling standards were perfected in various provinces, giving consumers an alternative to the original project. For many large retailers, this was a perfect solution in that they could sell an organic product and prominently label it so, without being forced to identify products containing GMOs. In general, though the Canadian system of biotechnology discussion relied on a relatively broad process of consultation, it tended to limit discussion to safety rather than expand it to issues of ethical concerns (Moore 2007). This alienated some of the participants and possibly undermined support for genetically modified products.

From the point of view of regulation, the federal government has tried to make the process friendlier to the approval of biotechnology. In 2003, the Framework for the Application of Precaution in Science-Based Decision Making and Risk was approved; although it mainly looked at establishing a precautionary principle for science and technology policy, it made it explicit that its application has limits. First, it is intended to be executed on a temporary basis (according to the progression of scientific knowledge). Second, domestic and international obligations might limit its application. Third, though public participation is welcome, its effective use is dependent on the time frame of the decision and the context. This means that the cost-benefit analysis to which the precautionary principle is subject involves both social and economic values.

Soon after the application of this framework, the federal government began working on the application of smart regulations to the field. The background work was based on the activity of the External Advisory Committee on Smart Regulation, which in 2004 released a final report including a recommendation for the streamlining of regulations along with the statement that the health and safety of Canadians must be protected by regulation. This encompassed the areas of biotechnology and environmental assessment. The principles on which the report argued for this streamlining were effectiveness, efficiency, timeliness, transparency, and accountability, but it also set forth four more claims that can be more easily contested. It called for a synchronization of Canadian policy with that of the United States; it claimed that risk assessment should be based on an instrumental cost-benefit analysis (much as the 2003 framework did); it noted that the private sector would easily be able to cooperate in the process of regulation; and it suggested that smart regulation is no regulation. To make matters more suspicious to some, the authors of the report tended to be drawn from a pro-business and pro-deregulation milieu.

Although many groups opposed this reading of the process of regulation, and there were some notable voices raised against this approach (Graham 2005), the government continued on this path of smart regulation, finally ending with the creation of the 2007 Cabinet Directive on Streamlining Regulation. Once again, consultation was involved, but it appears to have been heavily dominated by actors favourable to the smart regulation approach.[8] The directive notes the dual objectives of protecting Canadians while carefully examining the economic costs of doing so and the importance of carefully measuring the impacts of regulation on international competitiveness and international obligations before creating it.

Much the same approach continued to be followed with *Mobilizing Science and Technology to Canada's Advantage* (Canada 2007), the Science and Technology Policy backgrounder. In it, the federal government called for more private sector commitment to science and technology and the transformation of research into marketable products and services. This process is guided by four core principles: promoting world-class excellence, encouraging partnerships among actors, enhancing accountability of the system, and focusing on key priorities. These priorities are environmental science and technology, health and life sciences and technologies, natural resources and energy, and information and communication technologies.[9]

The new strategy also wrapped together the Advisory Council on Science and Technology, the Council of Science and Technology Advisors, and the Canadian Biotechnology Advisory Committee into the Science, Technology, and Innovation Council (STIC). However, as of March 2010, STIC had produced only a small innovation roadmap, presenting a set of subpriorities for the four priority areas and a report on the state of innovation in Canada for 2008.

The question remains why Canadian processes of consultation/engagement fail to have any great effect on the policy regime. One immediate answer can be gleaned from the scientific rationality principle at the root of the quasi-promotional regime. The principle drove the regulatory process toward favouring models of engagement rather than consultation. This is rooted in the attitude that sees the public as in need of "education" on issues of biotechnology, the argument being that, once educated, the public will respond better to innovation. Consumer ignorance and the level of engagement in public consultations also affected the attitude of the industry toward the adoption of biotechnology innovation. This is reflected in the comparative findings of Weldon and Laycock (2009) on the US, New

Zealand, and Canadian wine industries. In the Sonoma and New Zealand cases, where recent public consultations had raised the profile and controversy of biotechnology and GMOs, producers were concerned about adopting either biomarkers or GMOs, especially as first-wave innovators. In Canada, however, producers were much less concerned about possible responses from members of the public, who were considered unsophisticated regarding the application of biomarker technology, but were just as concerned as the American and New Zealand producers about the use of GMOs. The second answer lies in the commercialization approach of Canadian science and technology policies that, for some time now, have pushed toward conversion of the research into marketable products. Because of these elements, we believe that the role of participation in the Canadian context will, at least until some major changes are effected in core elements of the policy regime, remain bound to a voice option rather than a more participatory orientation.

NOTES

1 Participation of the public and stakeholder groups in the policy process when new technologies are involved has received a great deal of attention in recent years (Fischhoff and Fischhoff 2001; Haddow et al. 2007; Haga and Willard 2006; Metha 2004; Sharp, Yudell, and Wilson 2004; Tutton 2007).

2 The OECD BioTrack database is useful for a general comparison of approved biotechnology; see http://www2.oecd.org/.

3 Compare the Canadian approach with the EU regulations 1830/2003 on Traceability and Labelling of GMOs and 1829/2003 on Genetically Modified (GM) Food and Feed (implemented in 2004), which required that any more than a 0.9 percent of unintended presence of an EU-approved genetically engineered substance would trigger a mandatory labelling of the product as GMO. Even if this regulation exempted from labelling products such as milk, eggs, and meat from animals fed with GMO feeds, it created massive limitations on trade, and, in 2006, the World Trade Organization ruled that this was a de facto moratorium on US, Canadian, and Argentine products. General international standards have also been elusive; the Codex Committee on Food Labelling (CCFL) of the Codex Alimentarius Commission has discussed this topic for over fifteen years without making much progress.

4 Under this standard, processing aids, enzymes below 0.01 percent by weight in a food as offered for sale (for exceptions, see paragraph 6.2.7.a), veterinary biologics, animal feeds, and substrates for micro-organisms (where the substrate itself is not present in the finished food product) do not affect whether a food or ingredient is considered to be a product of genetic engineering.

5 The case *President and Fellows of Harvard College v. Canada (Commissioner of Patents)*, [2002] SCC 76 (the Harvard Mouse case), established that higher life forms

did not fall under the definition of invention found in Section 2 of the Patent Act: "any new and useful art, process, machine, manufacture or composition of matter, or any new and useful improvement in any art, process, machine, manufacture or composition of matter."

6 The format might not be the most reassuring for opponents of GMOs but certainly illustrates the relevance attached by the federal authorities to the scientific rationality principle. "Scientific questions or information will be forwarded to CFIA and Health Canada evaluators for consideration in the assessment. Non-scientific input will be evaluated and appropriate ways of addressing it will be explored." See http://www.inspection.gc.ca/.

7 On the topic of food from cloned animals, Canada appears to be more inclined toward a precautionary principle. An interim policy (the Food Directorate Interim Policy on Foods from Cloned Animals) was put forward in 2003 requesting a voluntary moratorium on the development of cloned animals, which is still in place, until more information emerged. A similar moratorium was put in place in 2001 in the United States by the Food and Drug Administration (FDA), but in January 2008, the FDA concluded that meat and milk from cloned animals are safe for human consumption. In July 2008, the European Food Safety Authority concluded that there was no evidence of any difference between cloned animals and regularly bred ones in terms of their health risk when used as food.

8 For a list of participants in this phase of consultation, see the Regulation Canada website at http://www.regulation.gc.ca/.

9 The rhetoric of the federal science and technology strategy speaks of building three advantages: a people advantage, a knowledge advantage, and an entrepreneurial advantage.

REFERENCES

Abelson, J., M. Giacomini, P. Lehoux, and F.P. Gauvin. 2007. "Bringing 'the Public' into Health Technology Assessment and Coverage Policy Decisions: From Principles to Practice." *Health Policy* 82, 1: 37-50.

Andrée, P. 2006. "An Analysis of Efforts to Improve Genetically Modified Food Regulation in Canada." *Science and Public Policy* 33, 3: 377-89.

Avard, D., G. Grégoire, and M.S. Jean. 2008. "Involving the Public in Public Health Genomics: A Review of Guidelines and Policy Statements." *GenEdit* 6, 2: 1-9.

Brown, J.L., and W. Qin. 2005. "Testing Public Policy Concepts to Inform Consumers about Genetically Engineered Foods." *Choices* 20, 4: 233-37.

Brunk, C.G. 2006. "Public Knowledge, Public Trust: Understanding the 'Knowledge Deficit.'" *Community Genetics* 9, 3: 178-83.

BSDE Expert Working Party. 2006. *BioPromise? Biotechnology, Sustainable Development, and Canada's Future Economy.* Report to the Canadian Biotechnology Advisory Committee. Ottawa: Canadian Biotechnology Advisory Committee.

Canada. 2001. *Action Plan of the Government of Canada in Response to the Royal Society of Canada Expert Panel Report.* Ottawa: Queen's Printer.

–. 2007. *Mobilizing Science and Technology to Canada's Advantage.* Ottawa: Industry Canada.

Cantley, M. 2007. *An Overview of Regulatory Tools and Frameworks for Modern Biotechnology: A Focus on Agro-Food*. Paris: OECD.

Castle, D., and K. Culver. 2006. "Public Engagement, Public Consultation, Innovation, and the Market." *Integrated Assessment Journal* 6, 2: 137-52.

Costa-Font, J., and E. Mossialos. 2005. "Is Dread of Genetically Modified Food Associated with the Consumers' Demand for Information?" *Applied Economics Letters* 12, 14: 859-63.

Coyle, F., and J. Fairweather. 2005. "Space, Time, and Nature: Exploring the Public Reception of Biotechnology in New Zealand." *Public Understanding of Science* 14, 2: 143-61.

de Jonge, J., J.C.M. van Trijpa, I.A. van der Lansa, R.J. Renesb, and L.J. Frewer. 2008. "How Trust in Institutions and Organizations Builds General Consumer Confidence in the Safety of Food: A Decomposition of Effects." *Appetite* 51, 2: 311-17.

Durant, R.F., and J.S. Legge. 2006. "'Wicked Problems,' Public Policy, and Administrative Theory: Lessons from the GM Food Regulatory Arena." *Administration and Society* 38, 3: 309-34.

Fischhoff, B., and I. Fischhoff. 2001. "Publics' Opinions about Biotechnologies." *Journal of Agrobiotechnology Management and Economics* 4: 155-62.

Graham, J.E. 2005. "Smart Regulation: Will the Government's Strategy Work?" *Canadian Medical Association Journal* 173, 12: 1469-70.

Greely, H.T. 2001. "Human Genomics Research: New Challenges for Research Ethics." *Perspectives in Biology and Medicine* 44, 2: 221-29.

Gutteling, J., L. Hanssen, N. van der Veer, and E. Seydel. 2006. "Trust in Governance and the Acceptance of Genetically Modified Food in the Netherlands." *Public Understanding of Science* 15, 1: 103-12.

Haddow, G., G. Laurie, S. Cunningham-Burley, and K.G. Hunter. 2007. "Tackling Community Concerns about Commercialisation and Genetic Research: A Modest Interdisciplinary Proposal." *Social Science and Medicine* 64, 2: 272-82.

Haga, S.B., and H.F. Willard. 2006. "Defining the Spectrum of Genome Policy." *Nature Reviews Genetics* 7, 12: 966-72.

Hirschman A.O. 1970. *Exit, Voice, and Loyalty: Responses to Decline in Firms, Organizations, and States*. Cambridge: Harvard University Press.

Hornig Priest, S. 2006. "The Public Opinion Climate for Gene Technologies in Canada and the United States: Competing Voices, Contrasting Frames." *Public Understanding of Science* 15, 1: 55-71.

Isaac, G.E. 2001. "Transatlantic Regulatory Regionalism." *AgBiotech Bulletin* 9, 8: 1-4.

Kleinman, D.L., A.J. Kinchy, and R. Autry 2009. "Local Variation or Global Convergence in Agricultural Biotechnology Policy? A Comparative Analysis." *Science and Public Policy* 36, 5: 361-71.

Lindner, L.F. 2008. "Regulating Food Safety: The Power of Alignment and Drive towards Convergence." *Innovation: The European Journal of Social Science Research* 21, 2: 133-43.

Longstaff, H., M. Burgess, and P. Lewis. 2006. "Comparing Methods of Ethical Review for Biotechnology Related Issues." *Health Law Review* 15, 1: 37-38.

Montpetit, É. 2005. "A Policy Network Explanation of Biotechnology Policy Differences between the United States and Canada." *Journal of Public Policy* 25, 3: 339-66.

Moore, E.A. 2007. "The New Agriculture: Genetically-Engineered Food in Canada." *Policy and Society* 26, 1: 31-48.

Mwale, P.N. 2006. "Societal Deliberation on Genetically Modified Maize in Southern Africa: The Debateness and Publicness of the Zambian National Consultation on Genetically Modified Maize Food Aid in 2002." *Public Understanding of Science* 15, 1: 89-102.

Newell, P. 2008. "Lost in Translation? Domesticating Global Policy on Genetically Modified Organisms: Comparing India and China." *Global Society* 22, 1: 115-36.

Paarlberg, R.L. 2000. *Governing the GM Crop Revolution: Policy Choices for Developing Countries.* Washington, DC: International Food Policy Research Institute.

Peters, H.P., J.T. Lang, M. Sawicka, and W.K. Hallman. 2007. "Culture and Technological Innovation: Impact of Institutional Trust and Appreciation of Nature on Attitudes towards Food Biotechnology in the USA and Germany." *International Journal of Public Opinion Research* 19, 2: 191-220.

Pew Initiative on Food and Biotechnology. 2006. *Grapes, Wine, and Biotechnology: The Application of Biotechnology – Issues, Opportunity, and Challenges.* Washington, DC: Pew Initiative on Food and Biotechnology.

Rothmayr, A.C., and F. Varone. 2009. "Direct Legislation in North America and Europe: Promoting or Restricting Biotechnology?" *Journal of Comparative Policy Analysis* 11, 4: 425-49.

Seifert, F. 2006. "Local Steps in an International Career: A Danish-Style Consensus Conference in Austria." *Public Understanding of Science* 15, 1: 73-88.

Sharp, R.R., M.A. Yudell, and S.H. Wilson. 2004. "Shaping Science Policy in the Age of Genomics." *Nature Reviews Genetics* 5: 1-6.

Sheingate, A.D. 2006. "Promotion versus Precaution: The Evolution of Biotechnology Policy in the United States." *British Journal of Political Science* 36, 2: 243-68.

Talukder, K., and J. Kuzma. 2008. "Evaluating Technology Oversight through Multiple Frameworks: A Case Study of Genetically Engineered Cotton in India." *Science and Public Policy* 35, 2: 121-38.

Teisl, M.F., S. Radas, and B. Roe. 2008. "Struggles in Optimal Labelling: How Different Consumers React to Various Labels for Genetically Modified Foods." *International Journal of Consumer Studies* 32, 5: 447-56.

Tutton R., 2007 "Constructing Participation in Genetic Databases Citizenship, Governance, and Ambivalence." *Science Technology and Human Values* 32, 2: 175-92.

Weldon, S., and D. Laycock. 2009. "Public Opinion and Biotechnological Innovation." *Policy and Society* 28, 4: 315-25.

World Health Organization (WHO). 2005. *Genetics, Genomics, and the Patenting of DNA: Review of Potential Implications for Health in Developing Countries.* Geneva: WHO.

Wynne, B. 2006. "Public Engagement as a Means of Restoring Public Trust in Science: Hitting the Notes, but Missing the Music?" *Community Genetics* 9, 3: 211-20.

PART 5

MODES OF AND EXPERIMENTS IN PARTICIPATION

12

Swimming with Salmon
The Use of Journalism to Public Engagement Initiatives on Emerging Biotechnologies

DAVID M. SECKO

Sequencing of the human genome has propelled an interesting phenomenon: a strong, often mandated, requirement to investigate the social and ethical ramifications of obtaining personal and genetic information (e.g., the study of GE³LS – *g*enomics and its related *e*thical, *e*conomic, *e*nvironmental, *l*egal, and *s*ocial aspects – issues in Canada). Although gaining wider recognition through the human genome project, this movement now permeates much of biotechnology and, in particular, genomics.[1]

Investigation of the social issues related to genomics draws strength from the realization that genomic information provides the opportunity to change the world in ways that might be difficult to reverse. This realization functions alongside a growing discontent with how elected officials, agents of the state, and broader social and economic activities match citizens' perceptions of the public interest (which is, of course, not unitary). Both issues are apparent, for example, in the debates over genetically modified food (Hart 2002). There is growing recognition that governments need to both engage and consult with people regarding the effects of biotechnology and genomics. However, we are equally faced with the challenge of how best to undertake public engagement initiatives to avoid reinforcing preconceived notions, allowing strong interests to capture the process, or perpetuating informational bias.

In this chapter, I focus on the information that any effort toward public engagement must cope with to help meet such challenges. Take, for example, engaging publics on the social and ethical implications of obtaining and studying the genetic code of salmon[2] (referred to generically as "salmon genomics"). Salmon are a popular food choice, sought by commercial and sport fishermen, of interest to scientists in terms of their evolution and genetics, of traditional importance to various Indigenous populations, and an important part of complex ecosystems. Salmon are also a contentious issue in terms of debates over aquaculture, fisheries management, transgenics, and the environment (Aerni 2004; Lackey 2003, 2005; Noakes et al. 2003). The design of a public engagement event or communication dealing with salmon genomics therefore needs to obtain and assess literature relevant to such debates, identify the viewpoints of stakeholders, and assemble a set of issues for discussion. In this case, much of the information will be scientific, and hence conceptually complex, and riddled with technical jargon and the vested interests of experts in developing their own fields. Furthermore, this information might be biased or not readily disclosed by stakeholders due to passionate feelings on the topic.

Efforts to engage publics about genomics therefore need to answer two questions.

- What information would someone need to effectively participate in democratic deliberation on a given topic?
- How does one obtain, assess, and present this information in a manner that encourages public engagement with the information and each other?

When considering how to obtain, assess, and present information in a media-saturated world, it is natural to think of *journalism*, an age-old craft adept at gathering and presenting understandable information to a wide audience. Moreover, when considering emerging biotechnologies in which information can be contentious and perhaps obstructed from view, it is perhaps more relevant to think of the theoretical arguments surrounding the role of journalism in a democracy, where, for example, journalism is an endeavour that at its best involves extended periods of discovery, appeals to our sense of justice, and a search for clarification or verification on versions of the "truth" (Kovach and Rosenstiel 2001).[3] In this frame, journalistic theory and practice can be thought of as tools to help provide a critical perspective on any information and, more importantly, any "reasons" put forward for public deliberation.

This chapter arises out of the recognition that journalism remains an underutilized tool in the effort to answer the above two questions on how best to cope with the informational needs of public engagement initiatives. This awareness has developed out of ongoing research to develop democratic tools for the effective engagement of publics in the governance of emerging biotechnologies in Canada (Burgess, O'Doherty, and Secko 2008; Secko, Burgess, and O'Doherty 2008). With this in mind, I provide some exploratory thinking on what the theory and practice of journalism can bring to public engagement initiatives, in part, by discussing how we have learned to swim while working to engage various publics in discussions of the social and ethical implications of salmon genomics.

Information, Public Engagement, and Emerging Biotechnologies

Actively Seeking Moral Perspectives
Public engagement initiatives draw from a substantial and complex body of literature. I cannot comprehensively review this wide variety of thinking here, which is addressed in other parts of this book (also see Abelson et al. 2003; Burgess and Tansey 2009; Felt and Fochler 2008; Rowe and Frewer 2005; Secko, Burgess, and O'Doherty 2008). It is useful, however, to briefly frame this chapter with my view on the need for public engagement.

So why engage citizens about genomics? As biotechnology has permeated society, the gene has risen to mythic prominence. Riding along with this celebrity status, broader social and political questions have layered onto the concept of the gene (Jasanoff 2005). Genomics has also risen in prominence within the biotechnology sector itself, where the effects of genomic knowledge on effective drug use and animal breeding are appearing. As such, there is merit to the argument that genomics can significantly affect our lives (Collins et al. 2003).

At times, emerging biotechnologies have been considered acceptable (often in the case of medical applications); at other times, such technologies have been controversial (most frequently in the case of genetically modified food). Questions have naturally arisen, therefore, over how best to govern the emergence of genomics technologies. For some, this has led to calls to democratize technology development (Feenberg 1999; Sclove 1999; Sherwin 2001) by asking the public to take on an important role in the emergence of new technologies. This view draws weight from the realization that experts and policy makers – that is, those who normally decide how to govern emerging genomic technologies – use norms or values in decision making that do not necessarily echo those of the public (Burgess and Tansey 2009).

This misalignment has importantly led some moral philosophers to suggest that we need to create ethical frameworks for governing biotechnology and, in particular, genomics (e.g., McDonald 2000; Sherwin 2001). One goal of these frameworks is to "*actively* seek out moral perspectives that help to identify and explore as many moral dimensions of the problem as possible" (Sherwin 2001, 11). Acquisition of these perspectives requires learning about viewpoints on a problem, and hence, in this case, democratizing technology development necessitates some form of "engagement" over genomics. However, many members of civil society have not yet become sufficiently interested in these issues to invest the time and resources required to understand the issues and try to influence policy. As a result, this wide understanding of the "public" is often underrepresented in representative democracy or public consultations.

Information to Support Public Engagement

There is a good deal of debate over the usefulness of public engagement initiatives as currently practised (Abelson et al. 2003; Secko, Burgess, and O'Doherty 2008; Webler, Tuler, and Krueger 2001). Public engagement has been critiqued, for example, for whether it simply reinforces preconceived notions or is easily captured by strong interests, whether it can be effectively turned into policy whose effects are defensible, whether its expense is justified when events can have limited reach or outputs that are ignored, and how endless events can cause public fatigue and trust issues. Some have gone so far as to suggest that public engagement initiatives shift the responsibilities of governments to citizen groups, thereby creating a form of "participation by tyranny" (Cooke and Kothari 2002).

From an informational perspective, a common critique of public engagement initiatives is that participants are not adequately informed and/or that the information that they do receive during an engagement process is biased. Nevertheless, though this is a challenge that must be met, to have participants meaningfully deliberate on complex scientific topics, there is still the need for accessible information.

I do not see this informing as an end goal, as the deficit model of public understanding of science might (Wynne 1993). I see accessible information as setting the scene for a more robust deliberative environment (O'Doherty, Burgess, and Secko 2010) – one where non-specialists can have confidence in providing input into complex biotechnological discussions and experts and stakeholders can have confidence that this input is based on the knowledge necessary for such deliberations.

MacLean and Burgess (2009) point to four components necessary to the process: two informational and two representational. The informational components involve providing, first, enough accessible scientific background so that participants understand the biotechnology in question and, second, contextual information on the social, legal, and ethical consequences of its use. In both cases, this is information that people are not expected to bring to a public engagement event. The representational components extend from this to provide, third, a diversity of cultural and personal perspectives on the topic that participants can use to judge their own experiences against and, fourth, the inclusion of underrepresented people to better reflect the range of people that a biotechnology will affect.

This production of suitable and accessible information for a public engagement event is not easy. For a contextually controversial topic such as salmon genomics, meeting these four components requires wading into a complex set of views in both public discourses and the expert literature. In considering this, it seemed natural to explore the role (if any) that the theory and practice of journalism can play in this information production as part of the wider goals of deliberative engagement – if for no other reason than that journalism is a powerful means of disseminating information and education in society.

However, there is also a normative argument to be made. It draws on one role of the science journalist in a democracy, namely that of a communicator who supports general scientific competency, and on the public's right to know (Lambeth 1992) and its ability to meaningfully interact with scientific discussions (Logan 2001). This normative argument acknowledges that science is a powerful force in shaping society and that a commitment to democracy therefore brings with it a need to engage people about the direction of science. In dealing with emerging genomic technologies that can be contentious and perhaps obstructed from view due to vested interests, public engagement initiatives cannot rely on generic information. They need critical and in-depth storytelling that explores the vested interests that characterize genomic science. Moreover, such democratic engagement requires clear information that is familiar, perhaps even comforting, to a wide variety of learning styles. Can journalism help to provide this?

Learning to Swim with Salmon

It was not out of thin air that the questions of whether and how journalism can more thoroughly support public engagement arose. Instead, as part of a larger, multi-disciplinary research team, I brought the perspective of a trained freelance science journalist. I see science journalism as having thick democratic roots (Logan 2001) and a theoretical purpose to provide publics with the information that they need to be "free and self-governing" (Kovach and Rosenstiel 2001, 17).[4] This expertise generated an opportunity to experiment with journalism as a supporting tool in our public engagement work on salmon genomics.

In this section, I showcase this supporting journalistic role, beginning with a short history of how it evolved out of our work in ethics, social science, and deliberative engagement, followed by how journalism can help to more systematically address a wider range of democratic deficits.

From Applied Ethics and Social Science to Deliberative Engagement and a Supporting Journalistic Role

Salmon genomics is a particularly interesting issue from a public engagement perspective. Current international efforts to sequence the Atlantic salmon genome (Davidson et al. 2009; Quinn et al. 2008) present a focal point between scientific innovation and a topic of cultural importance and political volatility. Moreover, prominent issues related to aquaculture, conservation, recreational fishing, and Aboriginal rights surround salmon (Lackey 2003), which, scientifically, is one of the most heavily studied fish on the planet. Salmon genomics, therefore, finds itself rooted in an atmosphere of economic and market pressures that can impact consumer choices and shape the application of genomic knowledge and technology, reflecting key characteristics of the problems for the democratic management of high technology.

Traditionally, professionals such as risk managers, policy experts, and public consultants were called on to investigate such issues. Indeed, our research team first approached public engagement from an applied ethics background, with ethical analysis proposed as a systematic means of identifying the range of public interests in a complex issue (Burgess 2004). However, such a role for ethics and ethics experts has been criticized for simply filling the gap left by the recognition that science and risk assessment are far from objective (Jasanoff 2005). Although ethics might be able to provide categories that describe the theoretical dimensions of ethical judgments, some suggest that it cannot authoritatively determine the "right" action. If issues related to biotechnology will not admit of clear and principled final answers, then any resolution or policy is inevitably political. Our team therefore turned to traditional social science research – including surveys, in-depth interviews, and focus groups – to deal with practical moral issues that ethics cannot resolve on theoretical grounds. We collected and analyzed empirical data on salmon genomics (Secko and Burgess, forthcoming; Secko, Burgess, and O'Doherty 2008) to help develop bodies of knowledge that map the issues of importance to a deliberating public (albeit in a way that might not always be readily digestible by a non-expert, again returning us to potential spaces for a supporting role for journalism).

Although the use of social science is meant to inform wider attempts at democratic engagement, such methods can decontextualize public opinion, creating potential divergences between what a participant says in a survey versus his or her conclusions if involved in a formal policy process. Irwin (2001) makes the point that *"social research framing"* often desires to capture citizens' views in a quick, relevant, and understandable manner, with close ties to institutional agendas. For Irwin, this is in contrast to a *deliberative democracy* model, which has affinities to approaches such as deliberative polling and citizen juries and grants a more active role to citizens in defining the issues to be discussed and the decisions to be made. Hence, if the desire was a genuine commitment to improving civic deliberation and the ability to shape policy decisions on controversial topics, we arrived at the need to undertake a deliberative model in our work on salmon genomics (O'Doherty, Burgess, and Secko 2010).

Each step in this short historical trajectory raised informational questions. Often, however, it was only the realization that with each step we were likely to want to distribute results to a wider public through journalists (which many research groups desire) that was considered in earnest. This is not unusual. Public engagement methods often deal with journalism simply as a way to disseminate findings and direct attention. In categorizing over 100 mechanisms of public engagement into three domains – (1) *public communication* (e.g., publicity, public hearings), (2) *public consultation* (e.g., opinion polls, focus groups), and (3) *public participation* (e.g., deliberative polling, citizen juries) – Rowe and Frewer (2005) only explicitly label the "public communication" category as involving interactions with journalism and only through distributing set information from official channels into media channels. This showcases how journalism has not been extensively considered in the public engagement literature as a practical approach that can support democratic engagement across Rowe and Frewer's three categories.

I would argue that ethics, deliberative democracy, and social science overlap in their objectives with the objectives and responsibilities of journalism, especially if journalism is viewed not as a stenographer but as a full participant – one with the skills to aggressively pursue information and explore vested interests, rigorously test them, and digestibly render this information to wide audiences. Ettema (2007, 143) puts it clearly that, with these skills, "journalism should be asked to participate not merely by presiding over an uncritical forum for reason-giving but by acting as a reasoning institution." This argument seeks to shift journalism from a method to disseminate the results of public engagement to a practical approach that can help to create democratic engagement itself.

In Theory: Journalism as a Tool to Support Engagement
In light of the roles that ethics and social science can play in democratic engagement, journalism can offer at least two additional things: a way to critically assess and render the information obtained by ethics, social science, and other methods for use in public engagement initiatives, and a way to reflect on and sustain democratic debate.

These points can be linked to a view of journalism as a particular form of knowledge generation that draws on a tradition of moral truth seeking, as can be seen in writings on investigative journalism (Ettema and Glasser 1998) or journalism ethics (Ward 2004). Views on what journalism is and is not (or theoretically should and should not be) are tremendously diverse, of course, and it is beyond my purpose here to delve deeply into these debates. Instead, it is more fruitful to consider what journalism as a tool to support engagement might entail.

Some time ago, Levine (1980) argued for the use of investigative reporting as a research method to understand social phenomena. His argument was set within the classic context of the methods used by Bernstein and Woodward to uncover the Watergate scandal. "Bernstein and Woodward's work deserves the accolade of art, but there was, in addition, distinct method to their approach. By making that method explicit, we may develop some further insight into qualitative research approaches" (626). The proposition was that such insight might come in the form of "field interviewing" techniques, which, in the case of many politically charged scientific debates, is no easy matter for a researcher.

I am more inclined to think of journalism in the context of this chapter as a practical approach to democratic engagement over a clearly defined research method. Levine (1980) nevertheless argued that a defined journalistic research method based on investigative reporting would involve (1) gathering hard (documents and records), soft (informants and interview data), and personal (intuitions and emotions) evidence; (2) theorizing on this evidence to give it social meaning; (3) checking the evidence against an editorial review process to ensure publication; and (4) publishing a story in a public manner, which is then confirmed, denied, or improved on by the authors or others.

Experiences working with salmon genomics suggest that a practical journalistic approach might be reframed to involve four components:

- *an assessment phase* that uses relevant literatures and observations to develop ideas on what type of information a particular public engagement initiative needs;
- *a viability pilot* to look at the obstacles, resources, and time needed to develop the required information and how it might be best rendered (print, visual, multi-media, etc.);
- *original journalistic research* that gathers Levine's hard, soft, and personal information on the topic in question and then utilizes an editorial review process to ensure the quality of evidence; and
- *production and evaluation,* in which compelling and understandable journalism is produced for use in a public engagement event, followed by its evaluation, for example, against MacLean and Burgess's (2009) informational and representational criteria internally and perhaps external vetting by relevant experts and stakeholders.

This systematic way of learning about a topic has caused Ho, Ho, and Ng (2006) to suggest that investigative journalism can be combined with social science into what they term "investigative research." This approach would be guided by an "understanding of the nature of the phenomena to be investigated, the questions raised, and the specific conditions of investigations" (31). Adding depth to the four components, investigative research would be (1) *disciplined* in requiring the investigator to have an attitude of reflexivity, (2) *naturalistic* in needing direct observations of the topic of interest, and (3) *in depth* in entailing continual and detailed examination of a topic without premature closure of investigation (Ho, Ho, and Ng 2006). The combination of these viewpoints traces how journalism can be a distinct tool bound by various social controls (e.g., editorial review, confirmation by others, reflexivity).

However, in contrast to some social science methods, journalism expects the investigator to become close to the subject and produce information that specifically draws out human interest aspects. As a result, journalism, with its "naturalistic" and morally compelling features, is particularly useful in terms of democratic engagement since it can provide rich but measured information. In particular, democratic engagement about biotechnology and genomics requires that such information be distilled from complex scientific debates, and, though ethics and social science can provide such information, journalism is already naturally well suited to generating it in a way that is inherently engaging but still supports the exploration of vested interests.

Despite this optimism, the skills that make a good journalist are rarely found in textbooks. Furthermore, how journalists and the media reflect expertise can enhance or undermine civic discourse. If experts are used as institutions that pronounce what is right or wrong, good or bad, policy, then the role of expertise and related journalism will be to abridge or suppress discussion among members of civil society to understand differences of opinion. On the other hand, using ethics experts alongside social scientists to identify the range of value considerations and social interests that is then supported by investigative journalism encourages reflection on the range of opinion and how to live cooperatively in a democracy. Accordingly, if we combine ethics, social science, deliberative engagement, and journalism, we begin to add aspects of reflective and sustained democratic debate to methods of engagement.

In Practice: Journalism as a Tool to Support Engagement

Returning to the example of engaging various publics in discussions of the social and ethical implications of salmon genomics, a goal was to create an environment in which a journalist (in this case, me) could support deliberative engagement in the fashion described theoretically above. A helpful feature in this regard was collaboration with an ongoing salmon genomics project (the consortium for Genomic Research on All Salmonids Project or cGRASP), which gave access to the scientific researchers, laboratories, and any information generated from the collaboration. The social scientists and ethicists working on the project equally acted as journalistic sources (Secko 2006). These sources were complemented by an interview study conducted with thirty-three international salmon genomics researchers (Secko and Burgess, forthcoming) as well as information from literature searches and conferences or workshops on topics related to salmon.

This exploratory work began the process of a journalistic approach to supporting the ability to map the issues of importance to a deliberating public. These issues (e.g., salmon and conservation genetics or perspectives on how salmon genomics relates to the sport-fishing industry) might not be compelling enough to draw a news organization into spending resources on a story in a newspaper or might not be found in the expert literature. One example in which the combination of hard (scientific papers), soft (scientist interviews), and personal (intuition on why interviewees avoided speaking about a topic) information led to subtleties used to inform a public engagement event was the issue of how transgenics related to salmon genomics. Transgenics is often discussed in the scientific literature in relation to how transgenic salmon can affect natural salmon populations (Aerni 2004), whereas focus groups with members of the public and experts point to a conflation of transgenics with salmon genomics (Tansey and Burgess 2008). It was clear through a *naturalistic* approach that some salmon genomic researchers thought that it was inappropriate to associate transgenic applications with research to sequence a salmon genome.

At first glance, it might seem obvious that researchers would not want to discuss a controversial technology such as transgenic salmon in a public forum. But this was not the case. Instead, journalistic interviews revealed that it was the framing of the issue as inherent to salmon genomics, as opposed to a contextual technology that could be undertaken regardless of whether a salmon genome was ever sequenced, that gave some researchers pause. On a human interest level, these researchers feared that transgenic salmon would easily overshadow other important issues (e.g., the use of genomics for the conservation of wild salmon or questions of Canadian commercial capacity versus foreign companies to make use of a salmon genome). They were aware that a salmon genome might help to create a transgenic fish in the future but that such information could be used to help assess the risks associated with the technology. Some discussion of transgenics could not be avoided in a public engagement event on salmon genomics, but some salmon genomics researchers needed it to be known that they were not pursuing transgenic applications in their work. How a deliberating public made use of this information was then up to these participants.

With a journalistic approach, such information obtained during investigation was presented in an accessible way. The most illustrative result of this process was the production of a twenty-six-page booklet used in a deliberative engagement event in November 2008 on sequencing the salmon genome (see http://salmongenetalk.com/). The booklet contained information on salmon genomics, genome sequencing, as well as perspectives on various contextual issues related to prominent areas of controversy surrounding salmon in British Columbia. The entire research team contributed information to the booklet, but its production was situated in an approach to journalism that emphasized critical storytelling, the exploration of vested interests that characterize salmon genome science, and a comfortable writing style equivalent to a Grade 10 reading level. However, just because this booklet was produced in this way does not mean that it effectively engaged people in democratic deliberation. Much work still needs to be done to experiment with the journalistic models used to create information for public engagement (see Secko 2007 for a pilot study on this matter).

Nonetheless, journalism is in a position to organize and assess the information necessary to help effectively engage people in democratic debate about salmon genomics. This connection between journalism and a research project does raise issues of bias and loss of perspective for the journalist and challenge traditional notions of objective reporting and the journalist's remaining free of commitments (cf. Ward 2004). However, though hidden biases are always a concern, many journalists have long challenged traditional notions of objective reporting, suggesting instead that they should report on matters as committed members of society. This has much in common with the goal to engage publics more effectively in democratic deliberation and suggests that the ultimate role of journalism is in need of further reflection.

Discussion: Journalism and Public Engagement

There is a lot being asked of journalism here. The profession is fractured, profit driven, suffering from a loss of advertising revenue, and often fiercely individualistic. Journalists and journalism scholars are also divided on the use and mission of journalism as related to its contributions to public policy and civic life. This chapter is only a start, and a complete assessment of the usefulness of including journalism in public engagement initiatives, and its capacity to contribute to informing and understanding perspectives, awaits completion.

However, there is room for optimism and growth. The use of journalism in our salmon genomics research suggests five broad areas for future inquiry into how journalism can contribute to new and more effective engagement processes.

1 Improving "mediated" deliberation (Page 1996): there is a long history of study of the news media's relationship to deliberation and how to sustain

debates (see Gastil 2008) but little work on how making journalists active participants in public engagement efforts can impact this pursuit.

2 Accountability: is there a role for journalists in holding public engagement initiatives accountable to their goals and deliberative ideals (see Ettema 2007)?

3 Effectiveness: can journalism help to facilitate the wider distribution and understanding of public engagement materials and thereby their reusability?

4 New forms of journalism: evolving new forms can enhance the ability of audiences to meaningfully engage with science (Secko 2007), perhaps through research partnerships between journalism schools and public engagement projects.

5 Development of methodology: the potential for refining a journalistic approach for presenting science and technology that demonstrates the relevance of a wide range of perspectives and aggressively pursues reasons (investigative research) for public engagement events.

By including journalists in the project of developing improved public engagement approaches, the resulting journalism (whatever its use) can better reflect the range of benefits, risks, and uncertainties and the range of perspectives arising in public dialogue. The practical limits on the number of participants in a deliberation emphasizes journalism's ability to inform the wider public about the processes and outcomes of public consultations and expert opinions, thereby extending the benefits of informed deliberation to a wider public. Furthermore, the evolution of new forms of journalism in this context might emphasize the rationale for informing and seeking public deliberation (e.g., tracking politicians' behaviour, identifying concerns related to an issue, educating about science, seeking convergence, or informing policy). People have limited time and resources and must be convinced that issues are sufficiently important to motivate learning about them. Explaining different views and seeking to establish reasonable common ground for policy can motivate increased engagement.

For most public policy, informing people of the actions of their elected officials and bureaucrats enables them to exercise political choice through elections and by voicing opinions to their representatives. But for issues in which there appears to be a democratic deficit, as described above, journalism needs to enhance citizens' abilities to knowledgeably discuss and engage different perspectives. Although research and public deliberation are pursued in a variety of venues, they are inevitably limited to small proportions of the population. But the valuable work of developing informative materials, understanding alternative perspectives, and seeking common ground can be more widely useful for journalists seeking to support an inquisitive and engaged readership. Firsthand knowledge of the challenges of informing and supporting respectful deliberation will also provide journalists with more specific knowledge about the potential audience for their work. This might help them not only to reflect on and sustain such democratic debates but also to justify their time and the risk that they might become (or be seen to become) too sympathetic to the cause. However, as Ettema (2007) has pointed out, the arguments here raise paradoxes for further reflection on how journalists can be seen as fair-minded moderators of information while also being committed participants.

It is also an open question whether journalists want to be involved. This is a potential limitation of the wide adoption of journalism as a means of supporting public engagement initiatives, especially with independence being a core value for many professional journalism organizations. Besley and Roberts (2009) recently interviewed nineteen US-based print journalists who had covered some aspect of a deliberative engagement event about their views on the process and its effectiveness. These journalists expressed many positive views on these "interesting experiments," but their interest in deliberative engagement seemed to be linked to the novelty of the process as opposed to a commitment to deliberative ideals or improving the ability of underrepresented groups to shape policy decisions on controversial topics. Yet, it is difficult to predict how individual journalists will react to requests to join a public engagement initiative in the above vein.

There is the wider question of whether increased support for journalists working with scientists, social scientists, humanists, and policy makers on issues that seem to be of importance to society will enhance the democratic mission of journalism. In general, journalism is recognized for its influence on civic dialogue and public policy. Although the mechanisms and impacts of journalism are debated, its influence is not. However, even if the notion of a social contract between journalists and society seems romantic, to the extent that journalists trade on the good favour and public duty of their sources, they incur a responsibility to use their privilege for social good. I have argued here that an important component of the social good produced by journalism is its potential contribution to deliberative democracy in the area of biotechnology. Although inability to fulfill a responsibility is a reason to relieve people of their duties, I have also shown that there is good reason to think that journalism is capable of making these contributions.

The final question is then how best to operationalize the responsibility for journalism to contribute to informed and deliberative dialogue that helps to address democratic deficits related to areas such as biotechnology. For now, I have described journalism as a tool to support engagement capable of effectively rendering reasons for public deliberation, but perhaps there is a deeper purpose for journalists to hold deliberants and institutions accountable to the very reasons that they have put forth.

NOTES

I would like to thank Michael M. Burgess for his helpful comments during the preparation of this chapter. This research was supported by the consortium for Genomics Research on All Salmonids Project (cGRASP), with funding from Genome Canada and Genome British Columbia, and by Genome Canada and Genome Quebec as part of the GE³LS component of the Genozymes for Bioproducts and Bioprocesses Development project.

1 Genomics can be defined as the study of genes and their functions and often involves mapping the entire genetic code of an organism and developing tools to analyze this information.
2 Salmon is the common name for several species of fish (e.g., coho, Chinook, chum, pink, and sockeye) that are born in fresh water, swim to the ocean, and then return to their natal streams to reproduce.
3 There is much ongoing debate, of course, about whether the practice of journalism at all resembles such democratic ideals and how, at its worst, journalism is a tool for propaganda of the powerful and an anti-democratic force that forsakes its responsibilities in pursuit of profit.
4 Again, it is much debated what this "information" should entail exactly and how it should be best obtained, interpreted, and presented. Furthermore, it is critical whether information, say over popular debate or holding institutions accountable, is what a democracy and its members really require. See Stromback (2005) for an interesting discussion on standards against which to judge this point.

REFERENCES

Abelson, J., P.G. Forest, J. Eyles, P. Smith, E. Martin, and F.P. Gauvin. 2003. "Deliberations about Deliberative Methods: Issues in the Design and Evaluation of Public Participation Processes." *Social Science and Medicine* 57, 2: 239-51.
Aerni, P. 2004. "Risk, Regulation, and Innovation: The Case of Aquaculture and Transgenic Fish." *Aquatic Sciences* 66, 3: 327-41.
Besley, J.C., and M.C. Roberts. 2010. "Qualitative Interviews with Journalists about Deliberative Public Engagement." *Journalism Practice* 4, 1: 66-81.

Burgess, M.M. 2004. "Public Consultation in Ethics: An Experiment in Representative Ethics." *Journal of Bioethical Inquiry* 1, 1: 4-13.

Burgess, M.M., K.C. O'Doherty, and D.M. Secko. 2008. "Biobanking in BC: Enhancing Discussions of the Future of Personalized Medicine through Deliberative Public Engagement." *Personalized Medicine* 5, 3: 285-96.

Burgess, M.M., and J. Tansey. 2009. "Technology, Democracy, and Ethics: Democratic Deficit and the Ethics of Public Engagement." In *Emerging Technologies: Hindsight and Foresight: Technology, Democracy, and Ethics,* edited by E. Einsiedel, 275-88. Vancouver: UBC Press.

Collins, F.S., E.D. Green, A.E. Guttmacher, and M.S. Guyer. 2003. "A Vision for the Future of Genomics Research." *Nature* 422: 835-47.

Cooke, B., and U. Kothari. 2002. *Participation: The New Tyranny.* London: Zed Books.

Davidson, W.S., Y. Guiguen, C.E. Rexroad 3rd, and S.W. Omholt. 2009. "Salmonid Genomic Sequencing Initiative: The Case for Sequencing the Genomes of Atlantic Salmon *(Salmo salar)* and Rainbow Trout *(Oncorynchus mykiss).*" http://www.cgrasp.org/.

Ettema, J. 2007. "Journalism as Reason-Giving: Deliberative Democracy, Institutional Accountability, and the News Media's Mission." *Political Communication* 24: 143-60.

Ettema, J.S., and T.L. Glasser. 1998. *Custodians of Conscience: Investigative Journalism and Public Virtue.* New York: Columbia University Press.

Feenberg, A. 1999. "Escaping the Iron Cage: Or, Subversive Rationalization and Democratic Theory." In *Democratising Technology,* edited by R. Von Schomberg, 1-15. Hengelo: International Centre for Human and Public Affairs.

Felt, U., and M. Fochler. 2008. "The Bottom-Up Meanings of the Concept of Public Participation in Science and Technology." *Science and Public Policy* 35, 7: 489-99.

Gastil, J. 2008. *Political Communication and Deliberation.* Thousand Oaks, CA: Sage.

Hart, K. 2002. *Eating in the Dark: America's Experiment with Genetically Modified Food.* New York: Pantheon.

Ho, D.Y.F., R.T.H. Ho, and S.M. Ng. 2006. "Investigative Research as a Knowledge-Generation Method: Discovering and Uncovering." *Journal for the Theory of Social Behavior* 36: 17-38.

Irwin, A. 2001. "Constructing the Scientific Citizen: Science and Democracy in the Biosciences." *Public Understanding of Science* 10: 1-18.

Jasanoff, S. 2005. *Designs on Nature: Science and Democracy in Europe and the United States.* Princeton: Princeton University Press.

Kovach, B., and T. Rosenstiel. 2001. *The Elements of Journalism.* New York: Three Rivers Press.

Lackey, R.T. 2003. "Pacific Northwest Salmon: Forecasting Their Status in 2100." *Reviews in Fisheries Science* 11, 1: 35-88.

–. 2005. "Economic Growth and Salmon Recovery: An Irreconcilable Conflict?" *Fisheries* 30, 3: 30-32.

Lambeth, E. 1992. *Committed Journalism: An Ethic for the Profession.* 2nd ed. Bloomington: Indiana University Press.

Levine, M. 1980. "Investigative Reporting as a Research Method: An Analysis of Bernstein and Woodward's *All the President's Men.*" *American Psychologist* 35: 626-38.

Logan, R. 2001. "Science Mass Communication: Its Conceptual History." *Science Communication* 23: 135-63.

MacLean S., and M. Burgess. 2010. "In the Public Interest: Assessing Expert and Stakeholder Influence in Public Deliberation about Biobanks." *Public Understanding of Science* 19, 4: 486-96.

McDonald, M. 2000. "Biotechnology, Ethics, and Government: A Synthesis." Ottawa: Canadian Biotechnology Advisory Committee.

Noakes, D.J., L. Fang, K.W. Hipel, and D.M Kilgour. 2003. "An Examination of the Salmon Aquaculture Conflict in British Columbia Using the Graph Model for Conflict Resolution." *Fisheries Management and Ecology* 10: 123-37.

O'Doherty, K., M. Burgess, and D.M. Secko. 2010. "Sequencing the Salmon Genome: A Deliberative Public Engagement." *Genomics, Society, and Policy* 6, 1: 16-33.

Page, B.I. 1996. *Who Deliberates? Mass Media in Modern Democracy.* Chicago: University of Chicago Press.

Quinn, N.L., et al. 2008. "Assessing the Feasibility of GS FLX Pyrosequencing for Sequencing the Atlantic Salmon Genome." *BMC Genomics* 9: 404

Rowe, G., and L.J. Frewer. 2005. "A Typology of Public Engagement Mechanisms." *Science, Technology, and Human Values* 30, 2: 251-90.

Sclove, R.E. 1999. "Design Criteria and Political Strategies for Democratizing Technology." In *Democratising Technology,* edited by R. Von Schomberg, 17-38. Hengelo: International Centre for Human and Public Affairs.

Secko, D.M. 2006. "What's Good, What's Bad, and Who Decides." *TREK Magazine,* November, 19-22.

—. 2007. "Learning to Swim with Salmon: Pilot Evaluation of Journalism as a Method to Create Information for Public Engagement." *Health Law Review* 15, 3: 32-35.

Secko, D.M., and M.M. Burgess. Forthcoming. "Assessing Moral Perspectives on the Technical Application of a Fish's DNA: An Interview Study with Salmon Genomic Researchers." In *Fishing and Farming Iconic Species in the Genomics Era: Cod and Salmon,* edited by K. Culver and K. O'Doherty. Ottawa: University of Ottawa Press.

Secko, D.M., M. Burgess, and K. O'Doherty. 2008. "Perspective on Engaging the Public in the Ethics of Emerging Biotechnologies: From Salmon to Biobanks to Neuroethics." *Accountability in Research* 15: 283-302.

Sherwin, S. 2001. "Towards an Adequate Ethical Framework for Setting Biotechnology Policy." Ottawa: Canadian Biotechnology Advisory Committee.

Stromback, J. 2005. "In Search of a Standard: Four Models of Democracy and Their Normative Implications for Journalism." *Journalism Studies* 3: 331-45.

Tansey J.D., and M. Burgess. 2008. "The Meanings of Genomics: A Focus Group Study of 'Interested' and Lay Classifications of Salmon Genomics." *Public Understanding of Science* 17: 473-84.

Ward, S.J.A. 2004. *The Invention of Journalism Ethics: The Path to Objectivity and Beyond.* Montreal: McGill-Queen's University Press.

Webler, T., S. Tuler, and R. Krueger. 2001. "What Is a Good Public Participation Process? Five Perspectives from the Public." *Environmental Management* 27, 3: 435-50.

Wynne, B. 1993. "Public Uptake of Science: A Case for Institutional Reflexivity." *Public Understanding of Science* 2, 4: 321-37.

Science Media and Public Participation
The Potential of Drama-Documentaries

GRACE REID

If... Cloning Could Cure Us (Morgan and Sutton 2004) was a British Broadcasting Corporation (BBC) media experiment that signified a unique shift in the public participation agenda. In the past, one of the main goals of BBC science programming was to improve the general public's scientific knowledge, but *If... Cloning Could Cure Us* went beyond this agenda in an effort to engage publics and give them a voice in scientific policy. For this reason, the program was short-listed for best factually based drama at the European Public Awareness of Science and Engineering TV Drama Festival in 2005.

If... Cloning Could Cure Us was originally broadcast to 750,000 viewers in December 2004 as part of the second season of the BBC's *IF* drama-documentary (dramadoc) series. Dramadocs typically dramatize historical events or lives of famous people, but they can also fuse dramatic and documentary conventions to "portray issues of concern to national or international communities in order to provoke discussion" (Paget 1998, 61). The dramadocs in the *IF* series were examples of this latter type of dramadoc. *If... Cloning Could Cure Us*, for instance, tackled therapeutic cloning as its issue of concern. Therapeutic cloning is currently a theoretical process that involves cloning a human embryo to extract stem cells, which can grow into organs and tissues to treat diseases (Rhind et al. 2003). The stem cell extraction process was once expected to destroy the cloned embryo, but Chung et al. (2008) developed an extraction technique that could leave the embryo alive. Although therapeutic cloning is illegal in some countries, the UK permits this type of research on embryos up to fourteen days old. However, there has yet to be any scientifically documented cases of successful therapeutic cloning procedures using human patients.

The hour-long *If... Cloning Could Cure Us* dramadoc was created to give audiences an opportunity to imagine the future benefits and risks associated with therapeutic cloning. The producers of *If... Cloning Could Cure Us* extrapolated existing knowledge of therapeutic cloning to create a courtroom drama set ten years in the future. The drama tells the story of a fictional scientist, Dr. Alex Douglas, charged with breaking UK law because she attempted to perform therapeutic cloning research on embryos older than fourteen days. During the trial, her lawyer argues that she should be exculpated for the crime under the defence of necessity. The lawyer maintains that Douglas *needed* to use nineteen-day embryos (which, unlike fourteen-day embryos, have begun to differentiate into spinal cord tissues) to save the life of her spinal injury patient, Andrew Holland. Several fictional witnesses are brought forward to testify for and against the defence of necessity. They include Douglas's colleague/former lover, who exposed her illegal research, and a maverick scientist who conducted unsuccessful human baby cloning experiments that led to tumour formations in his patients. The jury also hears from a Chechen woman who was allegedly exploited when she donated her eggs to similar research (i.e., she was paid very little to donate her eggs, and she had medical complications following the procedure to remove her eggs). Douglas and her patient also testify.

As the courtroom drama progresses, the narrative is interspersed with factual statements typed on the screen and documentary interviews with real-life experts (e.g., four natural scientists from a variety of fields, two ethicists, a "pro-life"[1] representative, and the chair of the UK Human Fertilisation and Embryology Authority). These documentary devices help audiences to interpret the fictional cloning scenario. At the end of the program, BBC viewers were asked to call in and vote on whether the fictional scientist, Douglas, should be found guilty or innocent of conducting therapeutic cloning research on embryos older than fourteen days. Voting cost ten pence per call, and the results of the phone poll determined whether the BBC aired the guilty or innocent ending that it had prepared in advance.

This chapter explores the ways in which the *If... Cloning Could Cure Us* dramadoc went further than the deficit agenda in its efforts to trigger public participation relating to therapeutic cloning. It begins by introducing the deficit approach to public participation and explains how it has evolved in the past decade. The chapter then describes the interview and focus group research methods used to investigate public participation associated with the dramadoc. Next, it presents the interview research findings by identifying three strategies that the dramadoc creators pursued to ensure that the program did more than simply inform and educate audiences about therapeutic cloning. The findings and discussion section also considers focus groups' responses to these strategies. The chapter concludes by reflecting on the dramadoc genre's potential for encouraging public participation.

Theoretical Framework

The International Association for Public Participation (IAP2 2007) places public participation activities on a continuum that flows from left to right, indicating activities that have increasing levels of impact on policy development (see Figure 13.1). On the far left of the continuum are activities that require minimum public involvement and therefore have minimum impact on public policies. Examples include websites or fact sheets that inform publics. Slightly to the right of these activities are public participation events that have the potential for more impact. They include events that consult with various publics to elicit feedback on a policy (e.g., surveys), followed by events that involve publics to ensure that their concerns are understood (e.g., workshops). Farther to the right are public participation endeavours that collaborate or partner with publics by incorporating their ideas into all parts of the process (e.g., citizen advisory committees). Finally, on the far right of the continuum are public participation pursuits such as delegated decisions that empower citizens by allowing them to determine the final outcome. This final type of public participation has the most significant impact on public policy development.

FIGURE 13.1

Public participation spectrum

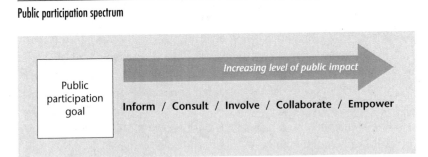

Source: Adapted with permission from http://www.iap2.org, copyright 2007 by IAP2.

In the mid-1980s, scientific public participation activities were typically situated on the left side of the IAP2 (2007) continuum because they emphasized educating and informing publics about science. This approach to public participation was put forward in a UK Royal Society report entitled *The Public Understanding of Science,* written by Sir Walter Bodmer in 1985. The report began by declaring a crisis in public interest in and support for science. It went on to claim that the crisis had been caused by the publics' low levels of scientific literacy. Bodmer therefore advocated the need for public participation activities that would enrich the publics' scientific knowledge. He believed that improvements in this knowledge would directly correlate to increases in public interest in and support for science. This approach to public participation has since been labelled the "deficit" model because it perceives publics as having deficits in scientific knowledge (Irwin 2001; Sturgis and Allum 2004).

A key outcome of the 1985 Royal Society report was the creation of the now obsolete Committee for Public Understanding of Science (COPUS), jointly established in 1986 by the Royal Society, the Royal Institution, and the British Association for the Advancement of Science (Miller 2001, 116). The committee promoted public participation activities that fostered public scientific literacy by educating publics about science. For example, it awarded prizes for science books that successfully conveyed science to various publics (Allan 2002, 53).

As the deficit movement took hold in the UK and spread to other parts of Europe, a similar movement was taking place in North America. In 1983, the influential US journal *Daedalus* dedicated an issue to concerns about the publics' low levels of scientific literacy (Turner 2008, 56). This resulted in institutions, such as the National Science Foundation, supporting public participation initiatives similar to those of COPUS. The Science Council of Canada (1984) quickly followed suit, and soon the deficit model was the dominant paradigm for public participation in science.

Less than a decade after its inception, however, cracks began to appear in the deficit approach to public participation. A number of studies challenged the fundamental assumptions behind the model. For example, Wynne (1992) found that people's lack of support was not necessarily caused by low levels of scientific literacy. In his research, Cumbrian sheep farmers were unwilling to support the expert advice of scientists in the aftermath of the Chernobyl radioactive fallout, not because they had low levels of scientific knowledge, but because they had alternative knowledge about sheep grazing that the scientists did not take into consideration. In 1995, Evans and

Durant published a study that challenged the deficit model's other assumption, that increased knowledge would consistently lead to increased support for scientific literacy. Their UK survey research showed that, though informed publics might be more inclined to support useful and basic areas of scientific research, they were less likely to support morally contentious areas of scientific research such as human embryology. In light of these findings, the UK government abandoned the deficit model in 2000 and endorsed what has now become known as the "contextualist" model of public participation (House of Lords 2000).

As the name suggests, the contextualist approach respects the context in which science is created and disseminated. This means that it acknowledges the idea that science is constructed, not to mention uncertain and "hotly contested" (Gregory and Miller 2001, 62). In light of the limitations of science, the contextualist approach recognizes the value of "other knowledge domains [e.g., lay knowledge] that influence attitudes towards science and technology in opposite or conflicting ways to factual scientific knowledge" (Sturgis and Allum 2004, 58). The model states that scientific policy development should take into consideration the ways that laypeople's needs, expectations, experiences, and cultures affect their attitudes toward science (Einsiedel 2004; Gross 1994; Sturgis and Allum 2004; Wynne 1992). For this reason, the contextualist framework encourages publics to actively contribute their unique knowledge to policy development through public participation activities that involve or collaborate with publics. This approach to public participation would be positioned in the centre or slightly to the right of centre on the IAP2 (2007) public participation spectrum because it encourages publics to impact policy development, but it does not believe that scientific policy making should be the responsibility of publics alone. The contextualist model sees policy development as a joint product of scientific and public knowledge (Gross 1994).

The past twenty years have seen a transformation in how public participation is practised. There has been a shift from activities that strive to get publics to understand science towards activities that strive to incorporate public understandings of science into scientific policy development. Publics have been asked to take a more active role in public participation and, therefore, have had more opportunities to impact policy outcomes. Consequently, public participation has progressed to a two-way process in which publics and scientists interact to develop scientific policy. Despite these achievements, Turner (2008) acknowledges that the contextualist model has still not replaced the initial deficit model of public participation. Irwin (2006)

says that many deficit activities continue to be practised under the erroneous label of contextualist public participation.

Research Methods

The research for this chapter consisted of nine interviews with the creators
of *If ... Cloning Could Cure Us* and twenty focus group screenings of the
program. Interviews were conducted with members at all levels of the *If ...
Cloning Could Cure Us* production team. Table 13.1 provides a complete list
of interview participants. During the one- or two-hour interviews, participants were asked questions about their roles in producing the dramadoc,
and they were encouraged to share their ideas about the type of public participation that they hoped the dramadoc would evoke (e.g., public participation that informed, consulted, involved, collaborated, or empowered
publics). These interviews were recorded, transcribed, and systematically
coded to identify producers' strategies for promoting public participation.

In addition to interviews, twenty focus groups were carried out to determine how 124 people in the UK and Canada responded to the dramadoc
text. Table 13.2 summarizes the types of focus groups conducted. The focus
groups were not intended to be representative of the general population;
instead, the variety and composition of focus groups were designed to elicit
an assortment of opinions about the *If ... Cloning Could Cure Us* program.
According to Kitzinger and Barbour (1999, 7), this approach to sampling

TABLE 13.1

Interview participants

Interviewee	Role in creating *If ... Cloning Could Cure Us*
1. Peter Barron	Editor for the first season of *IF*
2. Paul Woolwich	Executive producer for the second season of *IF*
3. Mary Downes	Series editor for the first and second season of *IF*
4. P.G. Morgan	Producer of *If ... Cloning Could Cure Us*
5. John Hay	Associate producer of *If ... Cloning Could Cure Us*
6. Jason Sutton	Scriptwriter for *If ... Cloning Could Cure Us*
7. Becky McCall	Researcher for *If ... Cloning Could Cure Us*
8. Dr. Stephen Minger	King's College London stem cell researcher and scientific consultant for *If ... Cloning Could Cure Us*
9. Jennifer Calvert	Actress who played the role of Dr. Alex Douglas

TABLE 13.2

Composition of focus groups

UK	Canada
General Public Groups (homogeneous)	
1 group with men	1 group with men
3 groups with women	1 group with women
Stakeholder Groups (homogeneous)	
2 groups with scientific experts	2 groups with scientific experts
• 1 with research scientists	• 1 with research scientists
• 1 with medical doctors	• 1 with medical doctors
2 groups with patients	2 groups with patients
• 1 with spinal injury patients	• 1 with spinal injury patients
• 1 with Parkinson's patients	• 1 with Parkinson's patients
2 groups with Catholics	2 groups with Catholics
Mixed Groups (heterogeneous)	
1 group composed of 2 members of the general public and 6 representatives from the above stakeholder groups	1 group composed of 2 members of the general public and 6 representatives from the above stakeholder groups

can lead to a wide range of responses, which provide greater insight into the research topic.

Strategies for recruiting participants varied with each type of focus group. Members of the general public, who had no vested interest in therapeutic cloning, were recruited largely by asking friends to provide contact details for their colleagues, relatives, and friends. Rotary Clubs also put me in touch with a few additional members of the general public. Scientists, on the other hand, were enlisted by sending cold call emails to scientists in bioscience or medical science departments at various universities. Doctors were found predominantly through medical mailing lists, whereas patients were contacted through patient organizations (i.e., the UK Spinal Injuries Association, the Canadian Paraplegic Society, the UK Parkinson's Disease Society, and the Parkinson's Society of Southern Alberta). Catholics were recruited through parish priests or by placing notices in parish bulletins.

At the start of the focus groups, participants were asked to complete a questionnaire before viewing *If... Cloning Could Cure Us.* The questionnaire was designed to assess participants' initial knowledge of the scientific process, ethical concerns, and legislation associated with therapeutic cloning. Participants then watched the *If ... Cloning Could Cure Us* dramadoc together. Afterward, they completed a post-viewing questionnaire to determine if and how the dramadoc affected their initial understanding of therapeutic cloning. Next, focus group participants spent an hour discussing the dramadoc. As the moderator, I initially encouraged participants to set the agenda for the discussion, but during the second half of the discussion I often prompted participants to discuss aspects of the program that they had overlooked. The focus group discussions were transcribed and systematically coded to identify consenting and dissenting views regarding *If ... Cloning Could Cure Us,* as recommended by Frankland and Bloor (1999). The following sections overview the findings of the interview and focus group research in relation to public participation. Although there were some variances in how different stakeholder groups responded to *If ... Cloning Could Cure Us,* there were no significant variances in how different nationalities or genders viewed the program.

Findings and Discussion

Television is often thought of as a deficit public participation activity because it involves one-way communication: science is conveyed to publics in the hope that it will increase their scientific literacy. In keeping with this agenda, the executive producer of the second season of *IF,* Paul Woolwich, identified *If ... Cloning Could Cure Us* "as a piece of television that was designed to inform and educate audiences" (interview with the author, 2007). As previously shown in Reid (2011), *If ... Cloning Could Cure Us* was largely successful in its efforts to inform and educate. An overwhelming 96 percent ($n = 119$) of focus group participants claimed during the focus group discussions that the dramadoc had presented them with new information, ideas, or opinions on the topic. Participants demonstrated specific examples of new knowledge in post-viewing questionnaires and focus group discussions. For example, many participants showed stronger understandings of the scientific processes, risks, ethical issues, and legal frameworks associated with therapeutic cloning. However, even after viewing the program, the majority of participants were unable to differentiate between therapeutic cloning and other forms of embryonic stem cell research (i.e., research that extracts stem cells from embryos that are left over from abortion or fertility procedures as opposed to cloned embryos). There was also some confusion about whether or not the UK's fourteen-day therapeutic cloning legislation was current. Despite failure to communicate these two key facts, *If ... Cloning Could Cure Us* did manage to significantly increase viewers' scientific literacy. Whereas Reid (2011) predominantly explored the ways that *If ... Cloning Could Cure Us* realized the deficit model of public participation, this chapter looks at three ways that the creators of the program managed to go further than promoting deficit-led public participation. The chapter also examines focus group participants' responses to these public participation efforts.

Employing a Fictional Narrative

Although the deficit approach to public participation focuses strictly on conveying scientific information, the creators of *If ... Cloning Could Cure Us* also incorporated a fictional narrative into the program. According to Paget (1998, 10), fiction is typically used in dramadocs to re-create a historical event or chronicle the life of a historical figure when no documentary camera footage exists. The *If ... Cloning Could Cure Us* production team, however, had other motivations for including a dramatic narrative in their program.

The creator of the *IF* series, Peter Barron, initially decided to make a program that fused documentary and drama conventions because he believed that a fictional storyline would make challenging topics more accessible to audiences who would not typically watch a pure documentary (interview with the author, 2007). In the case of *If ... Cloning Could Cure Us*, the production team thought that the drama portion of the program made science accessible by (1) camouflaging the fact that the program was predominantly intended to inform audiences about therapeutic cloning; (2) allowing viewers to appreciate the issue of therapeutic cloning through the eyes of the diverse characters in the drama; and (3) using dramatic plot devices such as love affairs and stunt work to stimulate and maintain audiences' interest in therapeutic cloning. The series editor, Mary Downes, said, "We worked extremely hard to genuinely dramatize real issues so that the audience would get engaged and really enjoy thinking about the topic of therapeutic cloning" (interview with the author, 2007).

In response to the production team's efforts to dramatize therapeutic cloning, the focus group research indicated that many participants enjoyed the fictional narrative. All twenty focus groups spent more than 65 percent of the discussion time talking about the dramatic storyline, as opposed to the documentary aspects of the program. Focus group participants were engaged and animated as they discussed the courtroom trial, though they did have several criticisms regarding the quality of the drama. Participants frequently used the fictional narrative to launch discussions about real issues surrounding therapeutic cloning. For example, the fictional Chechen character who was exploited for her eggs prompted lively discussions in fifteen of the twenty focus groups:

> *Cynthia:* But then you've got that woman from Chechnya in the show. It raises questions about the exploitation of egg donors from developing countries. That could be a bit disturbing ...
> *Kate:* They're not forcing them [to donate].
> *Roxy:* If someone wants to sell their eggs, that's fine.
> *Kate:* And the clinic didn't deliberately give the woman an infection [from the egg removal procedure].
> *Cynthia:* No, but the clinic should have helped her when she went back [with the infection].
> *Roxy:* Yes [they should have], but the clinic was in Chechnya, so it probably wasn't up to the [standard of] Western clinics.
> *Genevieve:* Yeah. They shouldn't be really using those clinics.
> *Kate:* The clinics over here should make sure they're checking these procedures to ensure that they are done ethically and properly. (Group 1, General Public)

This excerpt illustrates how the fictional narrative informed participants' understandings of the ethical issues associated with therapeutic cloning. It also provides evidence that disputes the deficit model's emphasis on solely communicating and understanding facts. Although understanding factual information is important, the focus group research indicated that fictional narratives tend to stimulate people's interest in scientific facts.

Critics of the dramadoc genre often worry that the fictional narrative will "manipulate," "mislead," or "confuse" audiences (Paget 1998, 1, 126; Rosenthal 1999, xix). However, in the case of *If ... Cloning Could Cure Us,* focus group participants were generally able to distinguish between fact and fiction. Nevertheless, the focus group research identified two significant instances of fact/fiction confusion that led to misunderstandings about therapeutic cloning. An earlier publication (i.e., Reid 2011) already discussed the first incident of fact/fiction confusion, which involved confusion over whether the UK's fourteen-day therapeutic cloning legislation was real.

The fictional narrative of *If ... Cloning Could Cure Us* is set in the year 2014, and Douglas is charged with "illegal experimentation on human embryos" under a legislative act supposedly enacted in 2008 (approximately two years after focus group participants saw the program). A fact screen and a brief expert interview in the documentary portion of the program clarify that the fictional legislation is already in operation in the UK under *real* 1990 legislation that makes it illegal to conduct research on embryos past the fourteen-day limit. However, at least 14 percent (*n* = 17) of focus group participants missed the clarification and consequently assumed that therapeutic cloning legislation had yet to be enacted in the UK.

When one participant asked a question about the fourteen-day legislation during a focus group discussion, another participant replied in a serious tone, "I guess you'll have to ask the legislators when they come around in 2008!" (Group 9, Catholics). Laughter followed this comment, but it became clear that nobody else in the group understood (or felt confident enough to clarify) that the legislation was in fact real. Confusion over the UK cloning legislation occurred in eleven of the twenty focus groups.

The assistant producer of *If ... Cloning Could Cure Us* was disappointed to learn that some focus group participants had been misled by the fictional 2008 legislation. In a 2007 interview, John Hay explained to me that the production team had decided to create the fictional legislation because the offence under the real 1990 legislation was somewhat vague and complicated (i.e., "keeping or using an embryo after the appearance of the primitive streak"). The team had therefore created the 2008 legislation with the clear-cut offence of "illegal experimentation on embryos" in an effort to ensure that audiences would understand precisely the criminal charge that Douglas faces. After hearing about the confusion, however, Hay acknowledged that creation of the 2008 legislation might have been unwise.

The second instance of fact/fiction confusion occurred when the fictional narrative in *If ... Cloning Could Cure Us* left several focus group participants with the misguided impression that the UK's fourteen-day legislation was in need of an *immediate* extension. During the courtroom trial, the defence lawyer claims that Douglas needs to conduct therapeutic cloning research on nineteen-day embryos (five days over the legal limit). The lawyer argues that nineteen-day embryos would eliminate the need to master the difficult differentiation process because day nineteen is when stem cells begin differentiating from cells that could turn into a variety of cell types to cells that are specifically designated to one area (e.g., nervous system cells to treat spinal injuries). At the time of the focus group research, however, scientists had only managed to grow a cloned embryo to day five (BBC 2005), which meant that there was no pressing need to extend the fourteen-day limit.

FIGURE 13.2

Question asked in post-viewing questionnaire: "If you answered that you are 'for' therapeutic cloning, would you like to see the law extended so that scientists can do research on embryos beyond the 14-day limit?"

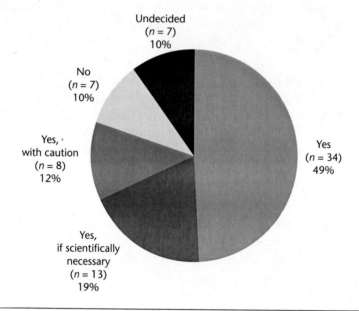

Nevertheless, many participants did not understand that the need to extend the fourteen-day legislation to nineteen days was a futuristic projection.

Most of the sixty-nine focus group participants who stated that they were "for" therapeutic cloning or "for with caution" also indicated in the post-viewing questionnaire that they would support extending the fourteen-day limit for experimentation on embryos (see Figure 13.2). Some participants (19 percent; n = 13) emphasized that they would support the extension only if it were "necessary" in order for therapeutic cloning to be successful. However, more than half of the participants, in favour of extending the fourteen-day limit, made no such stipulations. They were "for" the extension, or "for with caution," because they believed that an extension was essential to achieving success in the field of therapeutic cloning. In the post-viewing questionnaire, Anne wrote, "Yes, I would support extending the law because more time is needed to do this research" (Group 11, Spinal Injury Patients). Daniel also wrote, "Yes, as it might lead to the breakthrough" (Group 6, General Public). The idea that the fourteen-day limit currently needs to be extended, however, was actually a myth perpetuated by the *If ... Cloning Could Cure Us* program.

Focus group participants were unable to distinguish fact from fiction in the case of the extension because the program did not explicitly state that there was currently no reason to amend existing legislation. The documentary portion of the dramadoc, however, mentioned briefly that it was currently impossible to grow embryos to day fourteen. Participants could have used this knowledge to conclude that there were therefore no grounds for extending the current fourteen-day limit, but this was a lot to expect from mainly non-scientific participants presented with a large amount of new information. The few participants who understood that there was currently no reason to extend the existing limit for research on embryos tended to be doctors and scientists (though not all scientists and doctors recognized that the fourteen-day limit did not immediately need to be extended). The doctors and scientists who realized that there was no need for an extension were part of the group that voted "against" the extension. This finding suggests that people are not willing to support changes to scientific policy in anticipation of where science might lead; instead, they want evidence that policy changes are scientifically necessary.

The two instances of fact/fiction confusion in *If ... Cloning Could Cure Us* can provide advocates of the deficit model with ammunition for arguing that public participation endeavours need to stick to conveying facts. However, this confusion could easily have been avoided if producers had carried out audience research prior to broadcasting. Audience research might have involved prescreening the program to a few members of the general public and then asking them to participate in a brief discussion or questionnaire about the program. Any incidence of fact/fiction confusion could then have been rectified by adding a few extra lines of clarification in the documentary portion of the program.

Recognizing Lay Expertise

In addition to including a fictional narrative, the creators of *If ... Cloning Could Cure Us* challenged the deficit model of public participation when they decided to represent lay or non-scientific forms of expertise in the documentary portion of the program. The deficit model regards science as the only knowledge capable of informing debates. The contextualist model of public participation, on the other hand, respects scientific expertise but also tries to validate forms of non-scientific knowledge. The dramadoc producers' decision to conduct documentary interviews with a wide range of scientists and non-scientists was therefore more closely associated with the contextualist agenda. Downes, the *IF* series editor, said, "You need to give viewers an opportunity to listen to different viewpoints ... This means intelligently reflecting the spectrum of opinions across any topic" (interview with the author, 2007). In the case of *If ... Cloning Could Cure Us,* the spectrum of opinion consisted of four natural scientists (a stem cell scientist, an embryologist, a spinal injury researcher, and a Royal Society scientist) and four individuals representing other areas of expertise (two medical ethicists, a pro-life campaigner, and a policy maker). During the interviews, these experts were frequently called on to explain the facts behind the fictional scenario and offer their expert opinions.

Most focus group participants were pleased with the documentary interviews in *If ... Cloning Could Cure Us.* During all twenty focus group discussions, participants quoted almost verbatim both the scientific and non-scientific experts from the program. Although participants sometimes acknowledged that they had taken words or ideas from experts in the program, more often than not they passed ideas off as their own. Most participants used expert arguments to support their viewpoints, but occasionally people would reiterate an expert's argument that was contrary to their own positions and then explain why it was inadequate. This occurred in response to a comment that medical ethicist John Harris made during a documentary interview in *If ... Cloning Could Cure Us.* Harris stated that donors should be compensated for donating a tissue or organ:

> Everybody else is paid in transplantation. The nurses are paid. The surgeons are paid. The recipient is handsomely paid in kind – they get an organ. The only poor devil who is required to act altruistically is the donor, and that seems to me to be genuinely exploitative if one is looking for cases of exploitation.

One focus group participant understood but strongly disagreed with this line of reasoning. After viewing the dramadoc, James said:

> I thought it was very interesting that one of the contributors there, the chap with the beard, talked about how in organ donations the surgeons are being paid, and nurses are being paid, and so on. And that the only person being exploited is the donor, as if being a donor, as if giving, was subjecting yourself to exploitation! As if there was something not quite right about giving from the goodness of your heart! ... I don't know whether he intended to say

that, but I thought he sort of showed his true colours there, almost having contempt ... for the act of giving. (Group 9, Catholics)

During the focus groups, participants not only repeated arguments made by scientific and non-scientific experts but also discussed how the documentary interviews tried to explore issues from all sides. Claire said, "I thought that the show was well balanced by the commentators that were interviewed intermittently" (Group 19, Mixed Group). Her opinion was not shared by everyone, however. As a Catholic doctor who opposed therapeutic cloning, Angela thought that the expert interviews were biased in presenting the scientific community as uniformly in favour of therapeutic cloning (Group 20, Mixed Group).

The dramadoc interviews typically pitted scientists who supported therapeutic cloning against a pro-life representative and ethicist who opposed it. The notion that science and other forms of knowledge (e.g., ethics) are mutually exclusive is a common theme in media coverage of science (Priest 2001; Weingart, Muhl, and Pansegrau 2003, 258). However, as Angela pointed out, "There's a difference of opinion within the scientific community on this issue" (Group 20, Mixed Group). Her argument that the scientific community is not universally in favour of therapeutic cloning was supported by the fact that a number of the doctors and scientists who participated in the focus groups were against therapeutic cloning for a variety of reasons. In light of the disparity within the scientific community, Angela thought that the expert interviews should have involved scientists who were not in favour of therapeutic cloning. A couple of participants from other groups shared Angela's opinion. The inclusion of more diverse scientific opinions would have aligned *If ... Cloning Could Cure Us* more closely with contextualist public participation, which acknowledges the uncertain and contested nature of science. It also would have shown that science and other forms of expertise do not have to be at odds. The two spheres can work together to develop scientific policy, as the contextualist model of public participation recommends.

Involving Audiences in Policy Decisions

The most significant way that *If ... Cloning Could Cure Us* managed to transcend the deficit model of public participation was the phone vote, which challenged audiences to call in and decide whether Douglas should be found guilty or innocent of conducting illegal research on embryos older than fourteen days. Only 1.5 percent (n = 11,616) of viewers took part in the phone vote when the dramadoc was originally broadcast in 2004. The votes were calculated during a thirty-minute debate about stem cell research and therapeutic cloning, and the results showed that 81 percent (n = 9,381) of the people who called in voted in favour of seeing Douglas go free. Consequently, the BBC ran the innocent story ending for the dramadoc.

In a 2007 interview with me, *IF* Series Editor Downes explained the production team's reasons for including the interactive phone vote: "We wanted viewers to feel that they were in some way participating in the program rather than simply being preached to about the latest scientific research." This remark signalled the production team's intention to shift *If ... Cloning Could Cure Us* from solely a deficit public participation activity that informs audiences to a contextualist activity that actually involves audiences in making a decision about scientific policy. However, though the vote involved audiences in a question about a scientific policy, it is important to recognize that the policy question was fictional. For this reason, the vote in *If ... Cloning Could Cure Us* had little potential to impact real-life policy decisions. This type of public participation activity would therefore be situated to the left of centre on the IAP2 (2007) spectrum.

Despite the phone poll's limited potential for influencing policy, the voting question managed to stimulate discussion in the focus groups. Groups seemed to enjoy deliberating about the voting question. Most of the focus groups talked about it early in the discussion, without prompting from me as the moderator. Here is an extract illustrating how one focus group engaged with the voting question:

> *Jake:* I voted [that Douglas was] guilty, simply because her research on embryos was technically past the fourteen-day limit. And I don't think the defence established that her research was necessary to save the patient's life.
> *Daniel:* But then who established this fourteen-day rule? What if they were wrong?
> *Jake:* Yeah ... I'm not saying anything about where the law *should* be, but technically according to the law she is guilty.
> *Lewis:* Maybe the law should be changed.
> *Ethan:* Maybe, but it's not up to the jury to decide whether or not the law is right or not. It's just whether or not she broke the law. (Group 6, General Public)

Although participants seemed to share enthusiasm for conversing about the voting question, nine of the twenty focus groups raised concerns about the fictional nature of the question. Participants seemed to find it frustrating that the dramadoc spent an hour teaching them about therapeutic cloning, yet the vote did not allow them to apply their knowledge to a real-life policy scenario. Several participants wrote in the post-viewing questionnaire that they would not have called in to vote if they had seen the program when it was originally broadcast, because "the vote has no [real-world] consequence" (Group 14, Scientists).

Participants also worried that their vote on the fictional scenario might be misused to support real policy changes. Corrine said, "I didn't want to write down guilty because ... I thought that it would convey that I thought that stem research is wrong, which isn't the case. I just don't support the doctor's decision to carry out research past the fourteen-day limit" (Group 19, Mixed Group).

In another focus group, Nathan, who was in favour of therapeutic cloning, also expressed concern that his guilty vote would be misconstrued: "I would be worried that a headline might be splashed across the front page of the tabloid saying that 75 percent of the people voted that she was guilty and therefore did not support therapeutic cloning. This simply is not true" (Group 13, Parkinson's Patients).

These comments suggest that more participants would have been inclined to vote if the producers of *If ... Cloning Could Cure Us* had also included a voting question about a real-life policy issue (e.g., "are you for or against therapeutic cloning?"). The addition of this voting question would likely have encouraged voting by people who initially complained that the fictional vote did not have real-world consequences. It might also have persuaded people who were concerned about misuse of the fictional vote to participate, because a question based on reality could have clarified whether or not they supported therapeutic cloning. The addition of a real-world question would have moved *If ... Cloning Could Cure Us* farther to the right of the IAP2 public participation spectrum because the program could have had greater impact on scientific policy making. However, the BBC did create a forum on its website that offered audiences the opportunity to give their true opinions on stem cell research. Unfortunately, forum comments were posted only by a tiny number of people (.01 percent; $n = 39$) who had watched the original broadcast of the dramadoc.

Conclusion

The case study of *If ... Cloning Could Cure Us* highlights some of the successes that the dramadoc achieved in public participation, but it also suggests ways that the dramadoc genre can do more to encourage a contemporary approach to scientific public participation. In the future, dramadoc producers might want to consider conducting audience research prior to broadcasting to ensure that audiences fully benefit from the advantages of fictional narratives. They might also want to reflect on the value of acknowledging the uncertainty associated with scientific knowledge. Finally, dramadoc producers should think about combining their programs with devices such as phone polls or online discussions that offer audiences opportunities to contribute opinions on *real* scientific policy development. Such strategies will help the dramadoc genre realize its full potential as a contextualist mode of public participation.

NOTES

The data for this chapter were collected as part of my PhD in the School of Journalism, Media, and Cultural Studies at Cardiff University. My PhD was supervised by Dr. Jenny Kitzinger and funded by the UK Overseas Research Students Award Scheme; the UK Resource Centre for Women in Science, Engineering, and Technology; the Social Sciences and Humanities Research Council of Canada; and the Cardiff School of Journalism, Media, and Cultural Studies.

1　*Pro-life* is placed in quotation marks to acknowledge that pro-choice activists are trying to claim the term because they believe that their agenda also values life. As pro-choice activist Filipovic (2004, paragraph 1) said, "I think it's a travesty that 80,000 women die every year from unsafe abortions in countries where the procedure is illegal or highly regulated."

REFERENCES

Allan, S. 2002. *Media, Risk, and Science.* Buckingham: Open University Press.

Bodmer, W. 1985. *The Public Understanding of Science.* London: Royal Society.

British Broadcasting Corporation (BBC). 2005. "UK Scientists Clone Human Embryo." http://news.bbc.co.uk/.

Chung, Y., I. Klimanskaya, S. Becker, L. Tong, M. Maserati, S. Lu, T. Zdravkovic, I. Dusko, O. Genbacev, S. Fisher, A. Krtolica, and R. Lanza R. 2008. "Human Embryonic Stem Cell Lines Generated without Embryo Destruction." *Cell Stem Cell* 2, 2: 113-17.

Einsiedel, E.F. 2004. "Editorial." *Public Understanding of Science* 13, 1: 5-6.

Evans, G., and J. Durant. 1995. "The Relationship between Knowledge and Attitudes in the Public Understanding of Science in Britain." *Public Understanding of Science* 4, 1: 57-74.

Filipovic, J. 2004. "Reclaiming the Term 'Pro-Life.'" *The Oracle* (University of South Florida), 16 November. http://www.usoracle.com.

Frankland, J., and M. Bloor. 1999. "Some Issues Arising in the Systematic Analysis of Focus Group Materials." In *Developing Focus Group Research,* edited by R.S. Barbour and J. Kitzinger, 144-55. London: Sage Publications.

Gregory, J., and S. Miller. 2001. "Caught in the Crossfire: The Public's Role in the Science Wars." In *The One Culture? A Conversation about Science,* edited by J.A. Labinger and H. Collins, 61-72. Chicago: University of Chicago Press.

Gross, A.G. 1994. "The Roles of Rhetoric in Public Understanding of Science." *Public Understanding of Science* 3, 1: 3-23.

House of Lords. 2000. *Science and Society: Third Report.* London: Her Majesty's Stationery Office.

IAP2. 2007. "IAP2 Spectrum of Public Participation." http://www.iap2.org/.

Irwin, A. 2001. "Constructing the Scientific Citizen: Science and Democracy in the Biosciences." *Public Understanding of Science* 10, 1: 1-8.

–. 2006. "The Politics of Talk: Coming to Terms with the 'New' Scientific Governance." *Social Studies of Science* 36, 2: 299-320.

Kitzinger, J., and R.S. Barbour. 1999. "Introduction: The Challenge and Promise of Focus Groups." In *Developing Focus Group Research,* edited by R.S. Barbour and J. Kitzinger, 1-20. London: Sage Publications.

Miller, S. 2001. "Public Understanding of Science at the Crossroads." *Public Understanding of Science* 10, 1: 115-20.

Morgan, P.G. (producer), and J. Sutton (writer). 2004. *If ... Cloning Could Cure Us* [television series episode of *IF*]. London: BBC.

Paget, D. 1998. *No Other Way to Tell It: Dramadoc/Docudrama on Television.* Manchester: Manchester University Press.

Priest, S.H. 2001. *A Grain of Truth: The Media, the Public, and Biotechnology.* Oxford: Roman and Littlefield Publishers.

Reid, G. 2011. "The Television Drama-Documentary (Dramadoc) as a Form of Science Communication." *Public Understanding of Science* http://pus.sagepub.com/content/early/recent.

Rhind, S.M., et al. 2003. "Human Cloning: Can It Be Made Safe?" *Nature Reviews Genetics,* 4, 11: 855-64.

Rosenthal, A. 1999. "Introduction." In *Why Docudrama? Fact-Fiction on Film and TV,* edited by A. Rosenthal, xiii-xxi. Carbondale: Southern Illinois University Press.

Science Council of Canada. 1984. *Report 36: Science for Every Student – Educating Canadians for Tomorrow's World.* Ottawa: Science Council of Canada.

Sturgis, P., and N. Allum. 2004. "Science in Society: Re-evaluating the Deficit Model of Public Attitudes." *Public Understanding of Science* 13, 1: 55-74.

Turner, S. 2008. "School Science and Its Controversies, or, Whatever Happened to Scientific Literacy." *Public Understanding of Science* 17, 1: 55-72.

Weingart, P., C. Muhl, and P. Pansegrau. 2003. "Of Power Maniacs and Unethical Geniuses: Science and Scientists in Fiction Film." *Public Understanding of Science 12,* 4: 279-89.

Wynne, B. 1992. "Misunderstood Misunderstanding: Social Identities and Public Uptake of Science." *Public Understanding of Science* 1, 3: 281-304.

14

N-Reasons
Computer-Mediated Ethical Decision Support for Public Participation

PETER DANIELSON

The N-Reasons platform is designed as an explicitly normative polling instrument to improve public participation in ethically significant social decisions. Our research group (NERD)[1] designed N-Reasons to support both strong and weak experiments and used the results of both incrementally to improve the platform. In this chapter, we sketch a normative theory that motivates our claims that N-Reasons can improve *ethical* decision making and then show – using our own past work as a benchmark – that we can improve decision making. In particular, we will show that we can move from surveys to ethically significant decision procedures, endogenize expertise, overcome severe framing effects, and mitigate the primacy effect.

Normativity 101 with Some Help from Participants

Our goal is a democratic ethical oracle or moral compass that, given an ethically significant problem and a varied, modern population, outputs the ethically best option (or ethically ranked distribution of options). This is a very difficult goal; there is no generally accepted ethical theory that can provide decisive, substantive, practical advice.[2] In the face of persistent ethical disagreement at the substantive level, we join the retreat to proceduralism. In effect, we outsource substantive ethics to our participants. At the procedural level, there is wider agreement on the ethical appeal of the results of well-informed, uncoerced deliberation. In philosophical ethics, this approach is labelled contractarianism (Binmore 1994, 1998, 2004; Danielson 1998;

Gauthier 1991; Habermas 1984; Rawls 1971; Skyrms 1996) and in normative political science, deliberative democracy (Barabas 2004; Dryzek and List 2003; Fishkin 2005; Fishkin and Laslett 2003; List 2006). Although there are many theories of the social contract and even more of the best form of deliberative process, I will not engage in theoretical debate here. The focus on deliberative process operationalized by deliberative democrats suffices for my current task, first because we are committed to uncoerced, broadly inclusive, and reason-based public participation. Second, it advances our agenda, for leading deliberative democrats are committed to a practical, experimental approach.

From both the social contract and deliberative democracy, we take the idea of algorithmic social choice; from deliberative democracy, we take the idea of reasons. Legitimate social choices should be transparently democratic, and ethical choices should be supported by reasons. Unfortunately, these two demands are not easily satisfied together. Static surveys are relatively simple; it is easy to have each participant choose from among a small number of fixed alternatives (pro or con issues such as permitting online personal genomics or arming robotic aircraft) and, for the sake of transparency, report to him or her feedback on aggregate group choices. But once participants contribute the reasons for their choices, the set of thicker option-plus-reasons potentially expands exponentially.[3] Expanding choice in this way is deeply problematic. The problem is constructing a personal ethic e_i in relation to a complex social fact $[e_i \ldots e_n]$ and supporting decisions with appropriate reasons based on e_i. Expressed in informal game theory, equilibrium choice is normatively attractive but elusive: attractive because no one needs to choose in ignorance of what other participants choose. However, this is feasible only given the severely constrained options found in game theoretic models. Our normative challenge is to expand the scope of equilibrium choice to real problems, rich option sets, and agents with varied and socially interdependent reasons. We call this normative goal robust reflective equilibrium.

In effect, lacking a definitive ethical theory, we create a venue where citizens can bring their own to be factors in deciding issues. But the result might be a confusing and complex qualitative data set – a mess – rather than the transparent social decision to which we aspire.

Feasible Ethical Equilibrium: From Surveys to Social Decisions

We can put the point made normatively above in methodological terms. The NERD project used Internet surveys with Likert scale data augmented by free-form user comments (Ahmad, Bailey, and Danielson 2008; Ahmad et al. 2005; Ahmad et al. 2006; Danielson 2006, 2010a; Danielson et al. 2007; Danielson, Mesoudi, and Stanev 2008; Ilves et al. 2007; Mesoudi and Danielson 2007). Although unstructured commenting encourages participant input, it creates a double burden since, to play any role in the normative decision, comments need to be interpreted, and the authority of the qualitative interpreter is likely to be contested or at least is exogenous to the decision process.

Our new N-Reasons platform is based on this innovation: we encourage the participants themselves to think of reasons politically – that is, in the context of social decisions. First, we get them to classify their own comments as reasons for, against, or neutral to the option. By constraining choice in this way, we do more than encourage participants to classify their pre-existing comments; we "nudge" them to formulate a reason (Thaler and Sunstein 2009) and direct it to other participants.[4] Reasons are directed to other participants as part of the politics of social choice. Second, we encourage participants not to provide a new idiosyncratic reason if an existing one will do. That is, we encourage thinking of reasons as political devices, designed to gather support, mindful of the practical demands of feasible choice, namely that few participants are likely to read more than a handful of reasons both pro and con.

The result is a poll in which the options are others' reasons for or against the original substantive options. In effect, participants choose over a set of complex objects – with only a few options linked to many different reasons. This allows their choice to be nuanced (I choose No for this reason, not that one) but aggregative (nonetheless, I choose No, as did the majority).

Figure 14.1 shows the format of the second prototype N-Reasons design, the Personal Genomics and Privacy survey (Danielson and MacDonald 2009). The list of reasons is truncated to fit on one page. I will return to this figure to discuss several of its features below.

From Comments to Reasons

In this section, I consider the learning process that took us from optional, free-form comments to mandatory, constrained reasons. From the beginning, NERD surveys have asked for and collected text comments from participants. For most surveys, they were optional and private: that is, not published to other participants.

Public comments were introduced in the Parallel Ethical Worlds (PEW) survey (Mesoudi and Danielson 2007). Differences between public and

FIGURE 14.1

Personal genomics and privacy question 1

Question 1: Personal Genomics

1. "Personal genomics" companies (such as 23andMe, Navigenics, and deCode) offer at-home genetic tests that determine whether your genetic profile includes genetic markers statistically related to various diseases. (At-home services require customers to send a sample of saliva by courier; results are then returned either by mail or via a secure website.) These tests are based on relatively weak statistical correlations between genetic markers and diseases across large populations. Geneticists are doubtful about the usefulness of such tests, but the companies claim that they give customers insight into their own health and empower them by making them participants in their own healthcare.

If cost were not an issue, would you be interested in having your genetic profile assessed by a "personal genomics" company?

Please select the reasons closest to your own views on this question. You may also submit additional reasons. The most preferred are listed first. The popularity of each reason is shown after the text of the reason. (There have been a total of 68 votes on this topic so far.)

☐ **No** because I do not trust the confidentiality and privacy guarantees of for-profit companies (would the results be listed and sold to insurers or others)? I also would not trust that remnant specimens aren't being bio-banked and sold. I also have doubts about the usefulness of the tests ... **(lettuce) (expert reason)**   26%

☐ **Yes** because similar to the neutral advice of 'dearlizzie' but I would add one 'yes': building a community of 'could-be' patients may lead to global awareness about the usefulness of information and its related drawbacks, and counterbalance the perceived overwhelming power of medical doctors; for the curiosity aspect, and since I believe like many scientists that geographical origins have an impact on SNPs and genes, it may be interesting to check ancestry-related information along with genomic patterns. **(FXP) (expert reason)**   8%

☐ **Yes** because I would like to compare the results with facts about my family health history, and the results of my close relatives. This would help in making better health decisions. And yes, after thorough checking of privacy issues: although I would never trust the government, a private organization would also need to be verified by real users before I go for it. **(purple-er)**   4%

☐ **Yes** because it's my data **(lablogga)**   1%

☐ **No** because I'm concerned that the results could be used by insurance companies to deny coverage and/or affect future employment, should the information become public. **(bbwhalifax)**   2%

☐ **Neutral** because Yes, I am curious and confident about my ability to use the information properly; however, I am skeptical of privacy assuranes and would not want my information banked, sold or used against me by insurance firms. **(RoseC) (expert reason)**   2%

☐ **Neutral** because Do you mean "would I order a biochemical medical test from a company that claims not to provide medical advice?" Maybe. **(khhdocs) (expert reason)**   2%

☐ **No** because it is not anyone's business what my genetic code is. **(barryjbenjamin)**   2%

☐ **No** because The comment from Gensticists will skew the answers on the question ... as it is coercive Geneticists are doubtful about the usefulness of such tests. **(woodie) (expert reason)**   8%

☐ **Yes** because yes at a screening level only with anonymity. Confidentiality safeguards (i.e., number) to increase awareness of personal health issues for me and my family. Confirmatory testing to be done by government agency. **(bep) (expert reason)**   1%

Or add a new reason

○ Yes because
○ No because
○ Neutral because

Submit reason(s)

TABLE 14.1

Comment modalities

	Optional	Required
Private	All others	Animals in research: responsible conduct
Public	Parallel ethical worlds	N-Reasons

private commenting were noted in a contrast of the Neuroethics (Ahmad 2009) and Parallel Ethical Worlds surveys. These differences became the subject of an experiment in the Salmon Gene Talk survey (O'Doherty and Burgess 2011, Appendix B).

Figure 14.2 shows the length of comments (in characters) in the PEW and Neuroethics surveys. The PEW survey with public comments attracted significantly longer comments.[5] Of course, these results are only exploratory, for they compare two different sets of four questions on different topics. We designed an experiment to test the hypothesis that participants comment more in public by splitting the Salmon Gene Talk survey into two groups that differed only by the commenting condition. As in the PEW survey, the public commenters saw the comments of other members of the public group; the

FIGURE 14.2

Public versus private comments in two surveys

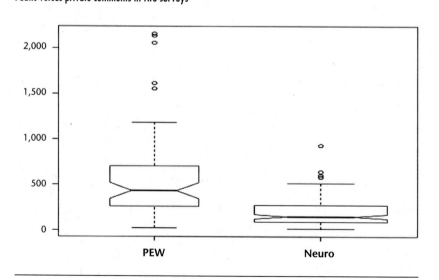

FIGURE 14.3

Private versus public comments in salmon gene talk survey

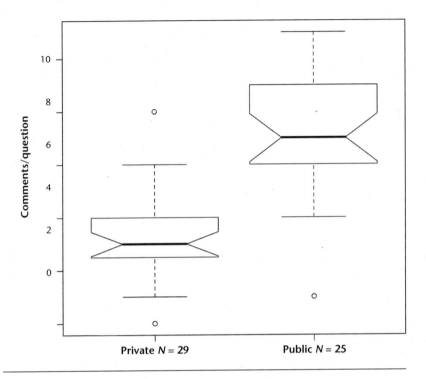

private commenters saw no comments. Other than amended instructions to cover publicity, the two groups saw the same set of twenty-seven questions.

The results shown in Figure 14.3 were striking. Public commenters made on average twice as many comments per question as private commenters. This speaks to concerns that publicity might deter participation; on the contrary, we found that publicity evidently motivated increased participation.

Public comments do put an additional burden on researchers: as user-generated content, they need to be monitored for compliance with Behavioural Research Ethics Boards (BREB) guidelines. Monitoring needs to be done in a timely and fairly continuous way, else the site slow down to wait for batch approval. That said, spam and flaming have not been problems with any of our N-Reasons surveys to date. One user who self-identified as a robot upset the expectations of some other users for seriousness appropriate to a survey about ethics, but this minor upset did not approach BREB constraints.

It does suggest that endogenizing reporting "inappropriate content" is a feature worth exploring and experimenting with.

Requiring comments never occurred to us in the early days of the NERD project; we were too wary of discouraging participants into the recruiting of whom we and our collaborators had put so much effort. We were mistaken and too cautious. Elisabeth Ormandy changed our minds dramatically with her innovative, contingency-based survey on Animals in Research: Responsible Conduct (Ormandy, Schuppli, and Weary 2009a, 2009b). She required users to provide a reason in the comment box, with the prompt "Please comment on why you support/do not support this research." She also recruited using Facebook with such success that she was not concerned to lose a few participants. Nonetheless, her drop-off rates were low, so we learned that we could make comments mandatory, opening the path for N-Reasons.

Structured Choice, Dynamic Choice, and Framing by Reasons

Structuring choice by reasons solves the explosion of rich normative content found in comments. But it raises a new problem: each participant takes, in effect, a different survey, framed by the reasons and choices of those who have preceded her.[6] Worse, if two groups take the "same" survey, then they see totally disjointed sets of reasons. We use this worst case as the basis for the first experiment run on the N-Reasons platform. We partitioned participants in the Robot Ethics survey into three groups of similar size who saw only those reasons contributed by their own group members (Danielson 2010b).[7]

The most important question is whether the extreme framing effect induced by reasoned choice will undermine agreement. Our prediction was that the three groups would choose the same options (Yes, No, or Neutral), though their reasons might differ. This prediction was confirmed by the results shown in Figure 14.4.

Endogenizing Expertise and Qualitative Analysis (DIY Coding)

Our successful NERD surveys on Genetic Testing, Salmon Genomics and Aquaculture, Forestry, and Animals in Research all provided optional access to expert advisers. Although supporting our claims that these surveys are educational and provide a good basis for ethically informed decisions, they are both very expensive to provide and expose us to the criticism that we bias our surveys by constructing our own advisers. N-Reasons exogenize expertise; by attracting expert and lay participants, who have proved willing to provide expert reasons, and then filtering these reasons by participant

FIGURE 14.4

Social decision 1 = Yes, 2 = Neutral, 3 = No, for three groups in Robot Ethics N-Reasons

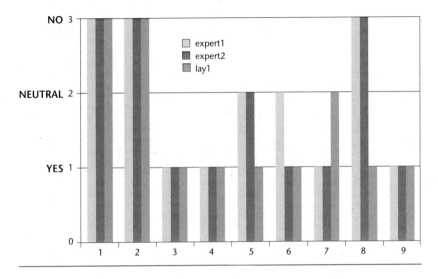

choice, well-regarded reasons are separated out for prominence (directly in the first N-Reasons survey on Robot Ethics). The quality of the top-ranked reasons has been remarkable and well-received by participants (as judged by exit comments) but not yet subjected to a more rigorous analysis. You can judge for yourself from the sample page in Figure 14.1.[8] Although we rely on participants to classify reasons, this process is not reliable; participants are not expert coders. For example, on the first question for the Personal Genomics and Privacy survey, we find

> *Yes* because I agree with the first no above but also because there is the potential for a huge economic drain if public/private resources [including manpower] are diverted to providing the follow-up services which I believe would eventually inevitably occur as a result of providing information. I am far more interested in diverting resources to help reduce poverty and improve literacy than I am about trying to outsmart mother nature. (Susan Chunick)

This provides an additional No reason but classifies itself as Yes, suggesting a role for an expert classifier. Or we can trust the group evaluation process to correct such errors, recalling that this reason appears so high on the page right now only because it is recent; it will drop down on the page unless others vote for it. Only experience will tell if the N-Reasons process manages to filter noisy participant input.

Multiple Votes

In response to participants' suggestions, and evident frustration with combining other participants' reasons, we modified the interface to allow multiple voting. If a participant votes for N reasons, each vote counts ⅟N. This is a clear case of incremental improvement based on experience if not experiment. There is an increase in complexity but no evidence so far that it is a problem. Likely, trying to work around the overly confining single vote rule puts more onerous demands on participants.

Dynamic Choice: Taming Primacy

In Danielson (2010b), we noted that the primacy effect was a problem of feasible choice. We argued there that we could, in principle, mitigate this effect by reordering the reasons displayed to participants. This left open the empirical question of whether this reordering would mitigate the primacy effect without, for example, confusing participants at the cost of transparency or even feasible choice. We put this question to the test by implementing the new display algorithm on two recent surveys.

Here is the explanation available to participants in the Personal Genomics survey:

> You vote for (one or more) reasons by selecting them. However, voting favors early reasons, which are shown to more voters and have "first mover advantage" attracting votes. To give later reasons a fair chance we have designed a display that favors recently selected reasons. We compute a score, measuring preference for reason R over other reasons on the page at the times when reason R was voted for. Reasons are ranked on the display by this score, not by the voting results that are indicated by the per cent (%) number following each reason. This preference score favors later reasons (for a short time), but only on the display. For reporting social decisions, votes are what counts.

We call this ranking, which blends recency and popularity, PrefDePref – or pdp.

FIGURE 14.5

Choices by reasons in two surveys

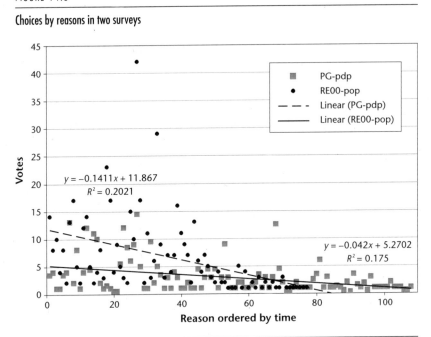

Figure 14.5 compares one (of the four) groups in the Robot Ethics survey, which displayed reasons by popularity, with the Personal Genomics survey, which displayed them using the pdp ranking just described. The groups were similarly sized ($n = 77$ and 108 reasons contributed in Robot Ethics [group = class 0, cohort 0] and Personal Genomics and Privacy, respectively). The trend lines are important; the popularity-based Robot Ethics survey has a much steeper trend: that is, earlier reasons had a much greater chance of being chosen. Under the display-ranking algorithm used in the Personal Genomics survey, designed to mitigate this primacy effect, the trend is much flatter (the popularity-line slope is 3.4 times the PrefDePref line slope). Of course, until we manage to get participants to return and reconsider their choices in the light of all reasons, early reasons will have an advantage (since they are the only reasons seen by early users). But we have managed to mitigate the additional advantage due to the primacy effect.

Finally, like the opening discussion of comments, these comparisons of surveys in terms of primacy are only weak experiments. A strong experiment, used for the framing effect, needs to be run between random samples taking the same survey.

Conclusion: What's Missing from N-Reasons?

We are not claiming that N-Reasons is an adequate platform for online democratic deliberation about the ethics of technology. We are claiming that it is better than NERD, which was better than some alternatives. We are also claiming that our incremental experimental methodology is a recipe for further improvement, without obvious limit. However, much is missing. Here are some of the missing elements that we can now see the need to prototype and test experimentally.

- How will N-Reasons work on larger groups?
- At what size of group does the reason-generation process saturate the space of reasons?
- Can direct comments on reasons be added (without confusion)?
- Generating options: Can the method be generalized from generating reasons (for fixed options) to generating options?

NOTES

This research was partially funded by Genome Canada through the offices of Genome British Columbia. Thanks to the NERD team, our collaborators in the surveys mentioned, and our participants for their enthusiastic support.

1 NERD is an acronym for Norms Evolving in Response to Dilemmas.
2 As Daniel Dennett (1998, 126) notes, "First, ... no ethical theory enjoys the near-universal acceptance of astronomy or meteorology, in spite of vigorous campaigns by the partisans. Second, there are no feasible algorithms or decision procedures for ethics as there are for celestial navigation. Third, the informal rules of thumb people actually use have never been actually derived from a background theory, but only guessed at, in an impressionistic derivation rather like that of our imagined meteorologists."
3 This follows from a diagonal argument, given that my choice for an option might reasonably be contingent on your reason for choosing it. See Danielson (1992, 219n7) for the details in the simplest two-agent–two-option case.
4 One surprising result of the NERD surveys was how often comments were directed, quite personally, to the survey authors. We provide a space for feedback to the authors at the end of N-Reasons surveys.
5 When the notches in the box plot do not overlap, this is "strong evidence" that the two medians differ (Chambers et al. 1983, 62).
6 This would not be a problem if participants would return and modify their votes in the light of new reasons. Unfortunately, we learned from the PEW experiment that

this almost never happens (4 out of approximately 400 opportunities). The PEW failure informed our goal of feasible equilibrium.

7   We partitioned the groups for the sake of this experiment, not because we believed that partitioning into small groups (around $n = 50$) was necessary to the success of the N-Reasons process. We are currently using the same Robot Ethics survey on larger groups to test this question.

8   Which is truncated to fit on the page. Full sets of reasons can be examined by visiting this survey and are, in this sense, self-documenting.

REFERENCES

Ahmad, R. 2009. "Investigating the Risks of Cognitive Enhancers." Paper presented at the Fourth International Conference on Applied Ethics, Sapporo, Japan.

Ahmad, R., et al. 2005. "Innovations in Web-Based Public Consultation: Is Public Opinion on Genomics Influenced by Social Feedback?" Paper presented at the First International Conference on e-Social Science, National Centre for e-Social Science, Manchester University.

Ahmad, R., J. Bailey, Z. Bornik, P. Danielson, H. Dowlatabadi, E. Levy, and H. Longstaff. 2006. "A Web-Based Instrument to Model Social Norms: NERD Design and Results." *Integrated Assessment* 6, 2: 9-36.

Ahmad, R., J. Bailey, and P. Danielson. 2008. "Analysis of an Innovative Survey Platform: Comparison of the Public's Responses to Human Health and Salmon Genomics Surveys." *Public Understanding of Science* 19, 2: 155-65.

Barabas, J. 2004. "How Deliberation Affects Policy Opinions." *American Political Science Review* 98, 4: 687-701.

Binmore, K. 1994. *Game Theory and the Social Contract: Playing Fair.* Cambridge, MA: MIT Press.

—. 1998. *Game Theory and the Social Contract: Just Playing.* Cambridge, MA: MIT Press.

—. 2004. *Natural Justice.* Cambridge, MA: MIT Press.

Chambers, J.M., W.S. Cleveland, B. Kleiner, and P.A. Turkey. 1983. *Graphical Methods for Data Analysis.* Belmont, CA: Wadsworth and Brooks/Cole.

Danielson, P. 1992. Artificial Morality: Virtuous Robots for Virtual Games. London: Routledge.

—. 1998. Evolution and the Social Contract. *Canadian Journal of Philosophy* 28, 4: 627-52.

—. 2006. *From Artificial Morality to NERD: Models, Experiments, and Robust Reflective Equilibrium.* Proceedings from Artificial Life 10: Achievements and Future Challenges for Artificial Life, Bloomington, IN.

—. 2010a. "A Collaborative Platform for Experiments in Ethics and Technology." In *Philosophy and Engineering: An Emerging Agenda,* edited by I. van de Poel and D. Goldberg, 239-52. Berlin: Springer.

—. 2010b. "Designing a Machine for Learning about the Ethics of Robotics: The N-Reasons Platform." *Robot Ethics and Human Ethics,* special issue of *Ethics and Information Technology* 12, 3: 251-61.

Danielson, P., R. Ahmad, Z. Bornik, H. Dowlatabadi, and E. Levy. 2007. "Deep, Cheap, and Improvable: Dynamic Democratic Norms and the Ethics of

Biotechnology." In *Ethics and the Life Sciences*, edited by F. Adams, 315-26. Charlottesville, VA: Philosophy Documentation Center.

Danielson, P., and C. MacDonald. 2009. "Personalized Ethics for Personalized Genomics." Paper presented at the 5th International DNA Sampling Conference, Banff, AB.

Danielson, P., A. Mesoudi, and R. Stanev. 2008. "NERD and Norms: Framework and Experiments." *Philosophy of Science* 75, 5: 830-42.

Dennett, D.C. 1989. "The Moral First Aid Manual." In *The Tanner Lectures on Human Values*, vol. 8, 121-47. Salt Lake City: University of Utah Press.

Dryzek, J.S., and C. List. 2003. "Social Choice Theory and Deliberative Democracy: A Reconciliation." *British Journal of Political Science* 33, 1: 1-28.

Fishkin, J.S. 2005. "Realizing Deliberative Democracy: Virtual and Face to Face Possibilities." *Journal of Zhejiang University (Humanities and Social Sciences)* 3, 1: 23.

Fishkin, J.S., and P. Laslett. 2003. *Debating Deliberative Democracy*. Malden, MA: Blackwell.

Gauthier, D. 1991. "Why Contractarianism?" In *Contractarianism and Rational Choice*, edited by P. Vallentyne, 15-30. New York: Cambridge University Press.

Habermas, J. 1984. *The Theory of Communicative Action*. Boston: Beacon Press.

Ilves, K.L., D.M. Secko, P.A. Danielson, and M. Burgess. 2007. "The Role of Public Consultation in Salmon Genomics Research: Exploration Using an Experimental Web-Based Survey Platform." Paper presented at the Genome BC Forum, Vancouver.

List, C. 2006. "The Discursive Dilemma and Public Reason." *Ethics* 116, 2: 362-402.

Mesoudi, A., and P. Danielson. 2007. "Parallel Ethical Worlds: An Experimental Design for Applied Ethics." Centre for Applied Ethics Working Paper. http://ge3ls-arch-working-papers/.

O'Doherty, K., and M. Burgess. *Sequencing the Salmon Genome: A Deliberative Public Engagement (Final Report)*. http://salmongenetalk.com/.

Ormandy, E., C. Schuppli, and D. Weary. 2009a. "An Interactive Web-Based Survey to Examine How People's Willingness to Accept Animal-Based Research Is Affected by Regulation." Paper presented at the Annual General Meeting of the Nordic Network on Agricultural and Environmental Ethics, Rakvere, Estonia.

–. 2009b. "Regulation Increases Public Acceptance of Animal-Based Research." Paper presented at the 7th World Congress on Alternatives and Animal Use in the Life Sciences, Rome.

Rawls, J. 1971. *A Theory of Justice*. Cambridge, MA: Harvard University Press.

Skyrms, B. 1996. *Evolution of the Social Contract*. Cambridge, UK: Cambridge University Press.

Thaler, R.H., and C.R. Sunstein. 2009. *Nudge: Improving Decisions about Health, Wealth, and Happiness*. New York: Penguin.

15 Contentious New Technology Introductions and e-Participation

KEITH CULVER

"Participation" is frequently invoked as a response to the problem of "disengagement." E-participation, using new information communication technologies (ICTs), has frequently been advanced as a particularly promising avenue to engagement. ICTs can provide tools to deliver more information of better quality to citizens at the time and place of their choosing, decreasing the cost of access to participation and promoting deliberative quality in a range of informal and formal processes. From the imagined gathering and automated integration of public opinion on GM food to the anticipated use of moving graphics and sound to explain to individuals just what a GMO is, proposed improvements to public consultation have been endlessly optimistic regarding the prospects for e-participation. Yet, progress has been halting and far from transformative in response to enduring problems of disengagement. What should we make of e-participation and its unfulfilled promise?

This chapter offers an argument in three steps. I first trace some of the factors driving the rise of e-participation, turning in the second part to provide a participant perspective on the institutional implementation of e-participation in the context of a Canadian provincial government. In light of that experience, I return in the third part to conceptual and practical questions regarding the goals and success conditions of participation exercises whose value is at least partly asserted to rest in their supporting democratic ideals and institutions, perhaps even to the point of transforming those institutions for the better. In this last part, I support a counterintuitive argument that the gap between promise and performance might be evidence not that e-participation has systematically been overrated but that its proper application domain has not yet arrived. The social demands of eco-innovation, I argue, might be the application domain where the solution of e-participation meets a problem that it is uniquely suited to face. Disengagement from deliberation regarding the introduction of novel technologies might usefully be met with border-crossing, collaborative, future-oriented e-participation applications whose operation can at last deliver a transformative and somewhat surprising kind of participation.

Legitimacy, Accountability, Consensus, and Social Cohesion

An embarrassingly short time ago, we were apparently at the dawn of a new way of life in Canada and other developed countries. The "end of history" claimed by Fukuyama met the "new economy" driven by new ICTs, leaving us ready for transformation into a "knowledge society." Or so the rhetoric went, taking shape in practice in various ways in various sectors and in various countries.

In Canada, we developed a particular kind of triple-helix system of innovation for the new economy (Etzkowitz and Leydesdorff 2000). Familiar elements include the relatively early development of the Network of Centres of Excellence system in the mid-1980s, inspired by computing networks, to imagine geographically dispersed yet intellectually united networks of researchers across Canada, working together and in concert with the private sector to translate knowledge into practice. As researchers across Canada sought new ways to work together, thinking about public infrastructure developed in parallel. Under Prime Minister Jean Chrétien in the late 1990s, the government of Canada supported plans to provide high-speed Internet access across the country – not just as a matter of allowing equal opportunity to send email and surf the web but also to enable a new era of distributed labour in which time zones and geographical distance would diminish in importance. And, as these research and infrastructural activities were pursued in the public and university sectors, the new economy's information technology private sector soared: from Research in Motion to Corel and Nortel, Canada was booming.

The influence, hope, and perhaps giddiness surrounding the new information technologies extended to hopes for social innovations in the operation of civil society organizations and the relations between citizens and the elected and unelected officials of our democratic institutions. If geography and time could be made less relevant to commerce, the hope went, then perhaps they could be made less relevant to democratic participation. If nothing else, some of us hoped, we might use electronic voting technology to defeat the Canadian problem of "seasonal voting" – our weather-driven need to hold elections while winter is at bay and polling stations are accessible without using snow shovels. Yet, some of us hoped for still more – an era of asynchronous politics in which political participation would occur from the place and at the time of citizens' choosing, contributing to the positive transformation of our democratic institutions. Optimistic figures said optimistic things – including, for example, the prediction that "E-democracy may be the 21st century's most seductive idea" (Culver 2003). We imagined that asynchronous politics might soon reach far beyond the easy starting point of e-voting and e-petitions, rapidly developing tools to enable efficient and wide-reaching deliberative practices fed by easily accessed knowledge (Culver 2004).

Public disengagement has been the object of much talk and action: as a subject of academic study, as a problem for retail politics, as a problem for public servants, as an opportunity for social innovators. Since my discussion here is driven by a participant perspective, I explore what I have seen and experienced as an academic standing from time to time on a bridge to practice, as a senior policy adviser to deputy minister-level actors in a series of e-participation or e-democracy efforts conducted by one provincial government. This perspective can be easily criticized as unrepresentative, narrow, biased, and so on. Or it can be recognized and evaluated for what it is – a participant perspective providing one experientially rich, albeit narrowly sourced, contribution to the gradual accumulation of case studies, surveys, and other evidence as we build our understanding of the merits of various kinds of public participation in the state's exercise of its claim to authority over social life. From a participant perspective on drivers of e-participation and its unfulfilled promise, it is natural to present a narrative in which specific drivers appear particularly salient and, in this narrative, as salient tensions. In that context, the engagement deficit appeared as a particular kind of easily measured, even if inchoate, sort of democratic deficit. The democratic deficit was easily measured in terms of declining voter turnout, even as the usual array of federally sponsored roundtables and other talking shops[1] identified a "disconnect" between politicians and citizens that runs rather deeper than the ballot box. As the disconnect was studied by academics, a few insights crossed the academy-to-government divide and made an impact. One useful example of a crossing point is found in a column written for *The Hill Times* by Sean Moore, then with Ottawa law firm Gowlings Lafleur Henderson. In this weekly newspaper, focused on events in Canada's federal Parliament and read by parliamentarians and policy wonks alike, he reported on the academic investigation of Professors Orsini and Phillips (2002) into citizens' involvement in policy networks.

Authored by Carleton University public administration professor Susan Phillips and political scientist Michael Orsini of York University's Glendon College, this is an interesting take on much of what ails government and politics today, not just in Ottawa but, as well, in provincial capitals and city halls across the land. It's about the disconnect between the governed and the governors.

Among their observations are the following:

- MPs (and MLAs) collect useful knowledge, but have little influence on policy
- Parliamentary committees are adversarial, lack resources, rely mainly on expert opinion, and have little impact
- Political parties are exclusionary and limit political discourse
- The public service relies too much on one-way communication
- Civil society organizations often have limited resources to participate fully and many are limited in their advocacy activities by government.

(Moore 2002)

A number of phrases used by Moore have become time worn, but their arrival at the time seemed fresh and penetrating, especially in talk of a "disconnect" between "the governed and the governors." The terms used to explain the nature of the disconnect are equally familiar: we hear of ineffective procedures that "collect" knowledge that ultimately has "but little influence" in an "adversarial" and "exclusionary" political environment inadequately informed by "one-way communication." The rhetoric of disconnection was likely made possible by citizens' undeniably significant declining interest in electoral politics, and that rhetoric gave rise to the discussion of new mechanisms of reconnection. Yet, the nature of that reconnection soon became complex – no simple "Rock the Vote" campaign starring a Canadian Madonna would suffice. As Moore (2002) relayed to parliamentarians and policy wonks, a reconnection would require a reimagination of the relationship between governed and governors, diminishing the hierarchical exercise of power in favour of something else:

Phillips and Orsini, in explaining what they observe as a renewed interest in citizen involvement in policy and government, see certain fundamental changes in the nature of governing and in civil society. Among them is what they and other observers have viewed as the shift from a top-down model of government to horizontal governance which is the process of governing by public policy networks including public, private and voluntary sector actors. "Whereas a traditional top-down approach emphasizes control and uniformity, horizontal governance recognizes that governments alone may not have the capacity, knowledge or legitimacy to solve complex public policy problems in a diverse society. Therefore it emphasizes collaboration and co-ordination."

There's a theme in there somewhere which I predict will turn up one way or another in someone's campaign-style speech in the months ahead.

It is easy to hear now-familiar tones of horizontality and network governance in this passage and the usual accompanying statement of motivation: assertion that the problems facing societies require solutions that governments lack. It is, I think, to the immense credit of Canada's parliamentarians and policy wonks that they did not sit idly by in the face of this challenge. Instead, they participated vigorously in near-government bodies and processes such as the Crossing Boundaries initiative led by Don Lenihan exploring the means of reconnection.

At this point, the opportunity for e-democracy tools seemed clear. When governments need citizens to be part of the solution to the problems faced by societies, and being part of the solution requires collaboration and co-ordination using shared knowledge, the new information communication technologies are inherently well suited to the challenge. The tools are many – from e-voting to online discussion forums, to information repositories at community and other scales, to collaborative interaction in simulated environments representing the consequences of particular policy choices. Much discussion of the tools occurred and a little experimentation too. Yet, eventually, as went Pets.com, so went e-democracy, morphing along the way into the current talk of e-participation, a rather feeble cousin. And, once again, a decade after the launch of the Crossing Boundaries project, we can see in its successor today, the Public Policy Forum's Public Engagement Program, the same motivations identified by Orsini and Phillips (Lenihan 2007). Little, it seems, has changed. The Public Policy Forum's website tells us that "a new generation of public processes is emerging that could transform the relationship between governments and their citizens. The Public Engagement Program has been established to raise awareness around the opportunities and issues this poses for governments and the public; and to help governments develop the knowledge, leadership and other skills they will need to make these processes work well."[2]

The new generation of processes noticeably lacks proposals to build on gains made in the use of e-democracy or e-participation techniques, and, instead, we hear mostly of new ways of organizing dialogue between governments and opinion leaders. Somewhere, it seems, between early days of heady optimism and the present day, e-democracy went missing. Why? In the next section, I distill some lessons from my experience as an adviser to a provincial government in the period 2002-7. It might be useful, however, first to offer a few observations about the political environment in which that experience occurred.

The passages quoted above are pieces of a zeitgeist in which old debates about the legitimacy of the state's claim to exclusive authority over social affairs ran into the New Public Management theory's emphasis on accountability, associated in various ways with parliamentarians' recognition of accountability rhetoric as a strategy for the stabilization of relations with skeptical voters. The partnership between legitimacy and accountability, even when used to transform a "governors and governed" relationship into a "representatives and represented" relationship, describes a fundamentally hierarchical relationship that even in the transformed version represents an asymmetry in final decision-making authority and proximity to knowledge and debate. These hierarchical relations are fundamentally in tension with the "collaboration and coordination" demands identified by Orsini and Phillips, often translated from these operational terms into the normative values of consensus and cohesion. It is possible, of course, to conceive of collaboration and coordination in ways that do not conflict with traditional hierarchical institutional structures involving governed-governor or representative-represented relations.

We can easily imagine, for example, situations in which federal, provincial, and territorial governments agree to coordinate the simultaneous operation of their specific powers to achieve some shared goal. Yet, these situations are typically the result of decisions already taken to be legitimated by their pedigree in institution-legitimating procedures such as elections, delegation of powers, and so on. These situations are not, then, instances of the sort of collaboration and coordination called for by Orsini and Phillips and many others.

Rather, collaboration and coordination, when converted to normative imperatives, address the inability of government to face problems alone, from identification of problems through identification of policy options and on to implementation of specific problem-solving or -mitigating options. This sort of coordination and collaboration is much more difficult – it is, as one senior public servant said to me, simply "not in the DNA of government." There is plenty of work on our political experience in this area, noting that consensus is both difficult for parliamentarians to find, and still more difficult to present to voters as the sort of win that merits re-election of a particular candidate for a particular party, whose difference from other parties has disappeared into the "everyone wins" ethos of consensus. Similarly, the goal of social cohesion, a kind of successful coordination of practical efforts, leaves parliamentarians looking less like representatives of competing visions of the good life and rather more like public servants – in short, like workers, not strategists.

So, a cynical conclusion might run, acceptance of a diminished governance role for Parliament, a role as convenor rather than governor, is simply not in the DNA of government. There is, therefore, no need to look far for the reasons that the tools of e-democracy and e-participation have not been widely taken up in Canada. Yet, this cynical conclusion would be premature, I think. A more plausible explanation is available, one that relies on a well-known situation facing new technological solutions: they need to meet the right problem, whose solution becomes their calling card and inspires the kind of investment needed to make full use of the technology's potential.

Distilled Experience

The lessons that I draw from my experience are closely connected to the starting points of the various exercises with which I was involved, all billed as web-based e-consultations with additional email brief-submission options and all either government run or requested by government and advertised as such. In each exercise that I conducted or advised with respect to structure and strategy, governments aimed to consult citizens with a view to actually making a decision leading to action on an issue in which new ideas were genuinely welcome. In terms of the OECD's distinction among information, participation, and active citizenship enabled by e-participation exercises, the activities within my experience fall between participation and active citizenship – an important classification if only to emphasize the role of these e-consultations as contributions to decision making rather than "tell and sell" exercises with respect to already adopted strategies or mere temperature taking to assess the public appetite for discussion of some issue (Castle and Culver 2006).

Each exercise in my experience was government conducted or government sponsored, yet in a collaborative environment. That environment involved my momentary adoption by governments from academic life as an adviser and use of private sector information technology firms to provide design and software writing or integration. Collaborative design and delivery have been driven by reasons including a lack of expertise within government, an attempt to use contract-based delivery of some services to control costs, and the straightforward addition of external labour to enable on-time delivery of an e-consultation beyond the ordinary labour capacity of government. Hosting e-consultations can be and has been carried out on government-based servers and servers operated by private sector partners. Debates regarding the hosting of e-consultations have typically revolved around technical issues such as reliability and normative issues such as compliance with privacy laws and, occasionally, clarity regarding responsibility for proper operation of the e-consultation.

A second dimension of collaboration has also been present throughout my experience of e-consultations, a matter of collaboration between online and offline consultation efforts. Ten years ago, when the Crossing Boundaries project began to bring attention to the possibility of improving the quality of democratic life in Canada by the use of novel information communication technologies, the decision to conduct e-consultations as complements or supplements to offline consultation was driven by often raised worries about the digital divide. As one City of Saint John councillor put it in an interview following an e-consultation there, "Regarding the Internet, the computer, or the website, there are some people who are slavish at them, but there are many people that didn't even know about them. And I'm wondering if maybe that alone was enough" (Culver and Howe 2004, 63). Although the substance of this worry diminished as time passed and e-commerce and e-government services were used more widely, concern regarding the merits of integrated off- and online approaches persisted. In part, the integrated or blended approach was driven by the absence of a strategic approach to the improvement of democratic processes – apart from Paul Martin's spoken worries about a "democratic deficit," there has been little political urgency driving experimentation with e-democracy and variants such as e-consulting and e-voting. Instead, we have seen an incremental and experimental approach in which e-democracy advocates in government have piggy-backed e-democracy techniques atop offline consultations on a variety of issues. We have not, in my experience, seen in any Canadian jurisdiction the formation of a politically supported, public service-delivered strategy for the integration of new technologies into democratic processes with a view to improving those processes. Instead, advocates have been opportunists, trialling e-democracy techniques wherever possible, in conjunction with offline approaches, often with sponsors hoping to reduce overall costs via the use of online techniques and at the same time, and possibly counterproductively, choosing conservatively among new technology options on the basis of familiarity and perceived risk.

In fairness to those advocates of e-democracy techniques who argued successfully for their trial, they often framed their projects as pilots whose success would justify scaling up their operation for broader deployment. Yet, even in pilot projects, some interesting cultural signals were evident. In piloting deliberation using online forums, for example, government officials with whom I worked were deeply concerned about what they framed as liability for illegal contributions – typically libel and advocacy of hatred against an identifiable group, as proscribed by the Criminal Code of Canada. Yet, behind this plausible concern, further concerns regarding the sources of knowledge provided during the consultation, and the mechanisms chosen for gathering citizens' opinions, indicated that beneath all spoken concerns lay familiar concerns predating the introduction of e-democracy techniques. Governments consulting publics regarding controversial issues are always sensitive to the sources of information presented to citizens as knowledge, particularly when governments are worried that misinformation might be introduced sheerly for political purposes. Governments have also long worried about loss of control – from town hall meetings degenerating into shouting matches to public consultations generating support for options running contrary to the principles of the government of the day. Pilot projects promoting citizen deliberation in online forums seem to strike some public servants as carrying all of the risks of a town hall meeting, with the added danger of electronic memory recording for all time what previously would have been one voice in a crowd, possibly recorded by a local reporter. Similarly, public officials have expressed concerns to me regarding the use of online surveys capable of being modified by a unique user as his or her opinions change throughout the course of a public consultation. Surely, I have been told, we can just get out there and find out what people think and be done with it – little good can come from an extended period of consultation

since it tends mostly to allow negative voices more time to organize disruptive interventions.

Much of the skepticism that I encountered was the expression of a particular kind of relation to the promise of e-democracy: a situation in which
there was a little to be gained by conducting an experiment in e-democracy
and only a little to be lost if the process went awry. Advocates of e-democracy
within governments have tended to be well informed regarding national
think-tank activities such as Crossing Boundaries and equally well informed
regarding the democratic deficit affecting Canada. Yet, these advocates have
operated within a public service that might easily be criticized for hiding
operational conservatism behind the convenient shield of deference to elected officials. As a consequence, policy issues available for trial e-consultation
have not been chosen because of any particular legitimacy deficit beyond
the capacity of government to face that gap without new tools. In short,
there has been no perceived *need* to experiment with e-democracy with
a further need to scale up from experiment to new standard practice.
E-democracy, in my experience, has always been trialled in situations in
which its merits have been difficult to distinguish from the accompanying
offline consultation or broader engagement practices.

Pulling these elements together, I have offered a picture of e-consultation
efforts in which a government conceiving of itself as having authoritative
control over a social situation collaborates with private sector technology
suppliers to deliver a democracy-enhancing service integrating with historical practice, in pilot projects treating readily available issues and not
issues particularly suited to, or perhaps in clear need of, e-consultation
tools. Viewed in this light, and with the benefit of hindsight, it is perhaps
unsurprising that so little of enduring effect has come out of the work
with which I have been associated. I now see in those efforts a series of errors. The largest and most fundamental, in my view, was the failure to pilot
e-consultation techniques in situations where there was a genuine need
for the special capabilities that they could bring. An associated failure was
the persistent managerialism and public service risk aversion that, under
the guise of pragmatism, selected the lowest-cost, and, not coincidentally,
the lowest-performing technologies available for what was then called path-
breaking experimentation in e-democracy. Although kids and adults have
played SimCity, and massively multi-player online role-playing games have
created three-dimensional worlds in Second Life, online consultation, now
sometimes independent of an offline counterpart, remains in the world of
web-based delivery of PDF consultation documents together with email

addresses for comments.[3] And, where the creators of Second Life and Sim City have at least recognized the need to market their new products to convince consumers of their merits, e-democracy activities in Canada have enjoyed an astonishingly low profile. So e-democracy has lost its lustre, and once again we hear from the inheritors of the Crossing Boundaries program that a new suite of tools promises a new beginning for relations between government and Canadians. Hence, the future for e-democracy looks a lot like recycling of the past.

E-Democracy for a Citified Future

After the gloom and skepticism of my preceding discussion, it might come as something of a surprise that I suspect that e-democracy has now, at last, met a problem that it is capable of solving, and – better yet – that problem needs e-democracy. The problem is the set of choices that we must make regarding the future of cities in the developed world. Cities pose a special challenge to policy choices and to democratic instruments chosen to inform those choices. As reported by the United Nations Population Information Network (www.un.org.popin/data.html), today just over 80 percent of the world's population dwells in cities, practically and intellectually multi-faceted entities concentrating people on geography, generating goods and services, and demanding water and energy and material resources to sustain human life and social practices, including transportation and construction in cities. As the world's population grows from today's 6.7 billion to a predicted 9.2 billion, total urban population is predicted to grow from 3.3 billion to 6.4 billion. Populations in the global north face a complex challenge – that of living in cities whose populations are aging, growing only modestly in numbers, on sites needing both remediation of environmental damage and subsequent adoption of resource-use practices that reduce the overall ecological footprint of cities. To reach a situation of sustainable development, cities, while remediating existing damage, must become more geographically compact to enable the more efficient operation of interwoven private and public transportation systems, including those that bring resources from afar to the concentrated needs of the city. Those transportation systems and housing and businesses must use renewable sources of energy more efficiently, in urban configurations whose impacts on surrounding ecosystems takes seriously the possibility that we might soon need what might be called "reconfigurable cities." They might be designed to be resilient in adapting to the uncertain consequences of forces such as climate change, immigration, and perhaps even pandemics.

As cities face the choices that they must make, they do so in a situation in which a vastly improved policy cycle can be enabled by the new information communication technologies, from information gathering, sharing, and interpretation through deliberation, choice of policy options, implementation of solutions, and monitoring of results. Recall from the first part of this chapter the conditions under which horizontal collaboration is needed across or among orders or levels of government – situations in which institutions of government face problems for which they lack the knowledge needed to solve them, lack the capacity to address them, or lack legitimacy as problem solvers. In Canada, for example, the situation of cities cries out for horizontal collaboration. Cities are the creatures of provincial law, substantially dependent on provincial politics and financial transfers, even while cities host the schools and universities governed by provincial educational law, the hospitals in which federal funding supports provincially administered services, the extensive urban built infrastructures that cannot be paid for by property taxes alone, and so on. Building the resilient eco-city of the future will require increased collaboration among all orders of government, public engagement, and information exchange.

The new ICTs are already fundamentally implicated in the steps taken by relatively independent government institutions in the context of the city – from embedding sensors for remote diagnosis of the structural health of civil infrastructure assets such as bridges, levees, and roads, to real-time air quality testing and traffic density monitoring, to advanced detectors monitoring smart grids of water and energy in real time with data storage allowing *post facto* analysis of trends and associations. The integration of these technologies enables significant efficiencies, reducing the environmental footprints of cities while simultaneously providing the information that citizens need to understand the impacts of public policies and their individual actions, to imagine and simulate new policies and actions, and to see the actual effects of public policy choices.

Consider these data integrated into 3D mapping enabled by new satellite-based geographical information systems, and imagine real-time data overlayed onto the 3D image – a city with a visible pulse. This is far from imaginary – Rostock, Germany, has a complete satellite-derived 3D picture of its buildings; Google Street View is developing precise street-level photographic coverage of streets in major cities, with 360 degree panorama capacity. With the addition of a game-like simulation interface, public officials can demonstrate the consequences of choices, and citizens can model the consequences of variations in air pollution, stress on energy and water grids, traffic, and their intersection – limited by the simulation, of course, but far less limited than at any time before. The importance of expert judgment in planning the future of cities is therefore emphasized rather than undermined.

Opportunities to take up these improved e-democracy techniques are fast approaching. Recent experiments in the use of reed beds to clean urban greywater discharge are now being considered for expansion into experiments in which the urban "heat island" effect might be mitigated by cooling evapotranspiration from the reed beds, whose use of urban space might be complemented by district biogas generation from waste and algal biofuel development. As these approaches are refined, transgenic microalgae are on the agenda for future biofuel generation experimentation, as are transgenic reeds for enhanced reed bed water treatment (especially for treatment of metals in greywater).

Here, at last, urban-dwelling citizens might have an urgent need to face the question of whether and how transgenic organisms will be accepted as ordinary working parts of their local environments. Although algae ponds might be relatively distant from urban housing, transgenic reed beds intended to both clean water and contribute to urban cooling will necessarily be located close to urban businesses, government offices, housing, and so on. This particular instance of urban biotechnology makes talk of transgenics far less abstract and distant. Urban biotechnology's local nature and great potential for contributing to urban cooling make the question of its adoption entirely concrete, real, here and now, potentially increasing the willingness of urban dwellers to use new ICT-based tools to understand proposed biotechnologies and negotiate their use or rejection.

As these activities might be developed, one significant problem encountered by first-generation e-democracy can be almost entirely avoided – that of the experimental nature of e-democracy projects piggybacked onto the "base" of offline public consultations and so typically underfunded as an optional extra. As the smart eco-city develops, electronic data *are* the base data – there are no maps needing to be scanned and put online, no analogue readings needing to be transcribed and put online. As the base data, these data are not specially commissioned prior to public consultation on a given issue and so are not a special extra cost beyond that of making them available in useful form to e-democracy designers and citizens. Conversion of raw data into useful analysis is a significant task, of course, but that task is aided significantly by the need of managers for simulation and visualization tools whose simpler versions can be used by citizens.

In the same vein, we might see a significant change to the old problem of piloting and scaling up, where pilots occurred frequently and scaling up more rarely. Scaling up depended significantly on data availability and citizen access to data in an era when the digital divide loomed large in concerns about inadvertent expansion of the gap between wealthy and poor via the creation of a special, information-privileged "digerati." The special situation of cities as relatively geographically compact social organizations has eased the gathering of information about their performance, and, though divides between wealthy and poor are enduring problems, digital divides tend to be least pronounced in urban areas. In cities, the highest-speed Internet access is supplied to government, businesses, and homes, typically routed to those geographical locations in ways that simultaneously lay the backbone of the intelligent city. High-speed Internet access is complemented by wi-fi access, paid or unpaid, mobile phone networks, and geographically accessible free Internet access in public facilities, including libraries, for those who cannot afford private access.

ICTs in cities do not stop at single-citizen devices, of course; the provision of information for deliberation can occur via public displays of information of various kinds in a range of public places. Train and subway stations can display real-time comparative energy consumption of currently operating rail versus road transportation, displays in city centres can communicate real-time data regarding airborne particulates, and traffic jam data whose communication to onboard computers can allow personal transportation of the future to be routed more quickly and efficiently – much as radio data systems integrated into GPS do in cars in many urban areas today. A final problem that I discussed, that of managerialism and risk aversion, will not, I think, be overcome via technology. It will be overcome in the same way that we will overcome the problems of capacity and legitimacy of governments as problem solvers.

In this city of the future, one of fully e-enabled deliberation, we might well find new public, private, and blended collaborative arrangements not imagined by previous generations of public officials. Or we might not. There is, notoriously, no guarantee that free people interpreting the same body of evidence will reach consensus on how to interpret that evidence. We might even find that increased deliberation produces more profound disagreements – deep wells of selfishness, for example, as NIMBY attitudes pit neighbourhood against neighbourhood. Fortunately, there are many ways to adjudicate these noisy and potentially fragmented debates; for example, neighbourhoods might be encouraged to use advanced simulation tools to take a longer view of their situations, changing the basis of evidence for debate. As techniques of this kind are deployed, on the backs of pervasive information systems leaving disagreement better informed than ever before, we might discover a transformed style of parliamentary representation, as members of Parliament become less representative and more expert and trusted negotiators. These negotiators can help us to live in a world in which the pressing demands of urban sustainability drive constant negotiation and renegotiation among interests and geographical zones of cities, a situation whose major danger, as long as the institutions of negotiation are strong and the resort to extralegal means is minimal, is not the collapse of civil society but the collapse of a particular kind of productivity as we spend more time in the agora and less time in the workshop. That, as countless country songs tell us, ain't all bad.

NOTES

1 For example, as noted in Chandler (2005, 5), "the Democratic Reform Secretariat of the Privy Council Office commissioned a consortium of Canadian thinktanks, including the Institute On Governance, to gather a range of stakeholders and experts in a variety of regions to discuss the nature of the democratic deficit in Canada. Over the course of five roundtable sessions, participants were asked to debate one of the following themes: the role of MPs (Halifax session); political parties (Montreal session); citizen consultation and engagement (Toronto session); Canada's evolving political culture and institutions (Calgary session); and democracy and the Canadian federation (Vancouver session). The Institute On Governance (IOG), a non-profit thinktank, led the roundtable at the Munk Centre for International Studies in Toronto which focussed on citizen engagement and consultation."

2 See http://www.ppforum.ca/.

3 A brief read of the government of Canada's "consulting Canadians" website is depressing indeed: the provision of information is still routinely confused with consultation, and use of the new information communication technologies appears to be intended largely as a paper-saving measure – consultations frequently amount to the posting of PDF files and an email address for comments. See http://www.consulting canadians.gc.ca/.

REFERENCES

Castle, D., and K. Culver. 2006. "Public Engagement, Public Consultation, Innovation, and the Market." *Integrated Assessment: Bridging Science and Policy* 6, 2: 137-52.

Chandler, J. 2005. *The Democratic Deficit, Citizen Engagement, and Consultation: A Roundtable Report*. Ottawa: Institute on Governance, 2005.

Culver, K. 2003. "The Future of e-Democracy: Lessons from Canada." *OpenDemocracy*, 13 November. http://www.opendemocracy.net/.

–. 2004. "How the New Information Communication Technologies Matter to the Theory of Law." *Canadian Journal of Law and Jurisprudence* 17, 2: 255-68.

Culver, K., and P. Howe. 2004. "Calling All Citizens: The Challenges of Public Consultation." *Canadian Journal of Public Administration* 47, 1: 52-75.

Etzkowitz, H., and L. Leydesdorff. 2000. "The Dynamics of Innovation: From National Systems and 'Mode 2' to a Triple Helix of University-Industry-Government Relations." *Research Policy* 29: 109-23.

Lenihan, D., ed. 2007. *Progressive Governance for Canadians.* Ottawa: Public Policy Forum.

Moore, S. 2002. "Confronting Canada's 'Democratic Deficit.'" *The Hill Times,* June.

Orsini, M., and S. Phillips. 2002. "Mapping the Links: Citizen Involvement in Policy Processes." Canadian Policy Research Networks Discussion Paper F 21. http://cprn.org/.

PART 6

UNDERSTANDING STAKEHOLDERS AND PUBLICS

16 Rethinking "Publics" and "Participation" in New Governance Contexts
Stakeholder Publics and Extended Forms of Participation

EDNA EINSIEDEL

As the new genetics has increasingly transformed traditional sectors, from food, crops, and livestock to industrial and medical products and their associated practices, frequently posed questions about risks and responsibilities, boundaries between nature and culture, and production and ownership of knowledge are also being reframed. In these contexts, ideas about publics and participation in these issues are similarly being reconfigured.

Earlier work on participation in environmental issues paved the way toward greater interest in dialogue with the public at the same time that organized social movements on women's issues, health, and consumerism were emerging. Contentious emerging technologies in the 1980s, along with governance challenges related to food safety and institutional trust, further encouraged explorations of "participatory technology assessments." Such approaches challenged the traditional domains of experts and marked a shift toward opening up science to society (Funtowicz and Ravetz 1993; Gibbons et al. 1994) as well as reformulating ideas about how publics were to be engaged. Much of the work on public participation in the 1980s and 1990s reflected these attempts at experimentation in public participation (Rowe and Frewer 2005). These efforts rested primarily on assumptions about the pre-eminence of state authority, the state's central role in determining technological trajectories, balanced by greater recognition of democratizing decision making by bringing publics back in; hence the unsurprising spotlight on public engagement efforts carried out by the state. A large variety of models of public participation focusing on "mini-publics" (Fung 2003) characterized these efforts (Rowe and Frewer 2005), the majority of which have been local or national in scope and focus.

At the same time, the arenas and institutions involved in many contentious issues around emerging technology have become increasingly global in scope. The case of technologies of genetic modification, for example, implicate international institutions (World Trade Organization, Food and Agricultural Organization [FAO], and other UN institutions), agreements, and outcomes for which national borders are not quite as meaningful and where the state is not always a central player (Buttel 1997). In these contexts, which publics are legitimate and meaningful participants in debates on technology? Which forms of "participation" count in assessing social influences on policy directions? The contexts of globalization have been marked by increasing numbers of civil society organizations, growing networked activities facilitated by the Internet and new social media, science and technology issues underpinned by international and national policy regimes, and more fluid boundaries for engagement by individuals and groups. These conditions have implications for how "publics" and "participation" are defined in the context of public participation.

Many scholars have argued over the term "public." As Habermas (1989, 13) maintained, "the use of 'public' and of 'the public' betrays a multiplicity of competing meanings." John Dewey (1927) suggested that there are many publics, each consisting of individuals who, together, are affected by a particular action or idea. Thus, each issue creates its own public, and each public will not normally consist of the same individuals who make up any other particular public, though every individual will, at any given time, be a member of many other publics. Just as policy makers employ constructions of "publics" who are invited to participate in formal processes of policy decision making (Einsiedel, Jones, and Brierley 2011), other publics emerge to disseminate their claims on technology debates.

These organized publics – typically referred to as members of stakeholder organizations – are distinguished from "the general public" or what Michael (2009) referred to as "publics in particular" and "publics in general." Participatory technology assessments have tried to focus on the latter group under the assumption that these are the voices not typically heard. Countries such as Canada might have two formal streams for participation, one for "the general public" and one for "multi-stakeholder groups," groups with recognized special interests. It is striking that much of the work on participatory technology assessment has focused on the former group. Separately, the literature on social movements has typically examined stakeholder activity in the interest of how such groups operate for social change; rarely are they discussed in the context of "public participation." This demarcation is evident in the association of groups with vested interests as symbolizing "much that deliberative procedures seek to overcome – partiality, competitiveness, and bargaining" (Hendriks 2006, 571). However, in the context of technology design and development, the idea of considering "relevant social groups" (Pinch and Bijker, 1987) with different interests in technological directions or ways of framing a given technology can provide the broader latitude needed to understand technological or policy outcomes. Case studies of mundane technologies such as the bicycle have demonstrated how particular groups of users, emphasizing values ranging from safety to masculinity or high fashion, can affect technological design and evolutionary paths at particular times (Pinch and Bijker 1997). This potential has been further demonstrated in the area of technological resistance. For instance, deaf people succeeded in impeding the further deployment of a cochlear implant (or "bionic ear," as its producers and the media labelled the product), partly because its designers had failed to recognize that deaf people form an established culture and community with its own norms and forms of communication (see Blume 1997).

In this chapter, I argue for the utility of broadening the discursive parameters for *publics* and *participation* when examining the engagement of multiple publics on emerging technologies such as biotechnology and genomics. Under this rubric, we can gain a better understanding of how "relevant social groups" influence technological developments and trajectories.

In describing two cases of civil society organizations participating in the public debates on genomics, I use ideas from the seminal work of McAdam, Tarrow, and Tilly (2001) on contentious politics and the *dynamics of contention*. They argue that different forms of contention – whether they relate to strikes, nationalism, revolution, or democratization – result from similar mechanisms and processes. They describe contentious politics as "episodic, public, collective interaction among makers of claims and their objects when (a) at least one government is a claimant, an object of claims, or a party to the claims and (b) the claims would, if realized, affect the interests of at least one of the claimants" (5).

Although the authors suggest the involvement of government as a key actor, they note instances of contention that involve non-governmental, formal, institutionalized power centres to which their framework can be applied. Understanding the role of the mechanisms that underlie social mobilization as a form of participation is key to elucidating the dynamics of contentious politics. They include *cognitive* mechanisms that demonstrate the significance of key "interpretive moments" in the political process through processes of social appropriation and interpretive attribution; *relational* mechanisms that describe the coalitional and organizational activities that make possible the extension of influences; and the *environmental* opportunities and threats that can account for the potential for social mobilization (see also Tilly and Tarrow 2006). These mechanisms of participation in policy debates that occur in broader public arenas are the focus of this chapter.

There are processes that the authors describe as relevant to each of these broad mechanisms, and I identify some of them in the two cases described below, though my analysis is not exhaustive. For example, relational mechanisms include those that "alter connections among people, groups, and interpersonal networks" (Tilly and Tarrow 2006, 25-26) as different groups establish connections, increase their interactions, and link with other established, recognized, or powerful others. A process of *certification* or *recertification* legitimates a position in the context of political competition by establishing a connection with valid political groups or powerful institutions. The environmental dimension includes the external occurrences that provide context to the subsequent mobilization activities. They are triggering events that shape subsequent dynamics of contention. A suddenly imposed grievance or experienced condition, a significant demographic change, and a scientific discovery illustrate occurrences that can trigger a spiral of opportunities. These environmental elements provide important contextual dimensions for explaining why certain actions are undertaken (or not). Cognitive mechanisms point to the processes of political perception, interpretation, and reconstruction, including the (re)interpretation of events as threat or opportunity, the creative use of social vocabularies to frame events, or the deployment of symbols and images to strategically shift meanings. These mechanisms can further contribute to a shift in the scale of influence and operation of organizations from more localized to national or transnational arenas, or from limited to more extensive and dense networks of influence (Tarrow and McAdam 2004).

I provide two examples of the work of stakeholder organizations – organized publics – around issues of genomics. If one is to make any claims about the participation of different publics in emerging technologies, a researcher could usefully take account of broader discursive arenas and participants who might propose alternative technological directions or contest dominant visions of technological forms. In describing the two case examples, I am essentially illustrating *mechanisms of participation* as portraying the dynamics of contention around genomic technologies. I end the chapter with a discussion of the implications for understanding technology designs and trajectories.

Case 1: The Ban on Terminator Technology

In March 1998, the US Patent Office granted US patent 5,723,765 on the "Control of Plant Gene Expression" to Delta and Pine Land Company and the US Department of Agriculture (Oliver et al. 1998). The patent describes a set of interacting genetic elements that allows the controlled expression of value-added traits or seed viability in a crop plant. A trait such as reproductive capacity can be turned off to produce sterile seeds. Control of the expression can rest with either the owner of the plant variety or the farmer who grows the variety.

For Rural Advancement Foundation International (RAFI), a nongovernmental organization tracking patent activity in agriculture since the 1980s (see Crucible Group 1994), this scientific development in patenting represented a new assault on farmers' rights, a major focus of the organization's mobilization efforts. RAFI (subsequently known as ETC or Action Group on Erosion, Technology, and Concentration) developed the concept of farmers' rights, meant to embody concerns about genetic erosion and the "gene drain" from south to north (Aoki and Luvai 2007). RAFI coined the term "terminator technology" to stress one of the applications of this technology. This application made it possible to grow crops with seeds that are viable when sold to the farmer but that become sterile in subsequent harvests. Farmers would then be unable to maintain a commercial variety from their own seed stocks and have to return to the seed provider. An earlier study by Richard Jefferson and his co-authors for the secretariat of the Convention on Biological Diversity (CBD) on the consequences of the new technology proposed the term "genetic use restriction technologies" or GURTs (see Jefferson et al. 1999), but terminator technology, a shorthand quickly taken up in the media, became more widely adopted. "Suicide seeds" and "zombie seeds" were other popular labels for variants of this technology.

When word got out about the first patent in 1998, RAFI and its allies launched a highly visible campaign against the technology (RAFI 1999), waged around the claim that it would prevent subsistence farmers from saving seeds and that pollen from the plants might sterilize neighbouring fields (see also Aoki and Luvai 2007). The competing representation from industry regarding the traditional frame of farming based on seed saving and sharing practices was that the centuries-old practice of farmer-saved seed is "a gross disadvantage to Third World farmers who inadvertently become locked into obsolete varieties because of their taking the 'easy road' and not planting newer, more productive varieties" (Steinbrecher and Mooney 1998, 177).

Not long after RAFI and its allies started their campaign, the world's largest non-profit agricultural research group and most influential agricultural research body in the south, the Consultative Group on International Agricultural Research (CGIAR), linked up with this effort and pledged never to use the technology in its crops. Its policy against the use of terminator technology declared the following:

> The CGIAR will not incorporate into its breeding materials any genetic systems designed to prevent seed germination. This is in recognition of (a) concerns over potential risks of its inadvertent or unintended spread through pollen; (b) the possibilities of sale or exchange of viable seed for planting; (c) the importance of farm-saved seed, particularly to resource-poor farmers; (d) potential negative impacts on genetic diversity; and (e) the importance of farmer selection and breeding for sustainable agriculture. (CGIAR 1998, 3)

This stance was echoed by the United Nations Food and Agricultural Organization's Panel of Eminent Experts on Ethics in Food and Agriculture, which stated that "the 'terminator seeds' generally are unethical" and found "it unacceptable to market seeds, the offspring of which a farmer cannot use again because the seeds could not germinate" (FAO 2001). In 2000, the media reported the following statement from the director general of the FAO, Dr. Jacques Diouf: "We are against [terminator genes]. We are happy to see that in the end some of the main multinationals which have been involved in implementing these terminator genes have decided to backtrack" (Reuters 2000). In the face of heated protests, Monsanto (now part of Pharmacia) similarly declared a moratorium on using the technology in 1999 when it was considering buying Delta Pine and Land (ETC Group 2006).

The efforts to ban terminator technology were particularly assisted by the Convention on Biological Diversity (CBD), enacted by the United

Nations as a legally binding framework agreement for the conservation and sustainable use of biological diversity and fair and equitable sharing of benefits arising from the use of biodiversity. Adopted in Rio de Janeiro, Brazil, in 1992, a moratorium on terminator technology was included in the CBD (Decision V/5, III).

Attempts were mounted to lift the moratorium in 2004 and again in 2006. Some governments, including those of Argentina, Australia, and Canada, headed the efforts to adopt a case-by-case approach to the use of terminator technologies, a move to which local media were alerted by domestic stakeholder groups (see, e.g., Bueckert 2006). A large international mobilization effort was mounted. In the end, the moratorium was upheld and strengthened at this CBD meeting. This outcome would not have been possible without the continuous vigilance and monitoring of the issue by a broad coalition of stakeholder organizations, embodied through the Internet-based "Ban Terminator" campaign. This campaign has 495 organizations listed as supporters, from countries ranging from Argentina to Zambia (Ban Terminator 2007).

The issues around this technology, as outlined on the campaign website (www.banterminator.org), include claims such as: "Terminator is a direct assault on the traditional knowledge, practices and the cultural life-ways of Indigenous peoples and rural communities across the world." The network links this claim to the United Nation's Ad Hoc Technical Expert Group report on the potential impacts of genetic use restriction technologies on small-holder farmers and Indigenous and local communities. It also claims that "Terminator is designed to stop farmers from saving and exchanging seed." Calculations by the ETC Group are presented, showing that, "if Terminator were commercialised, the extra seed costs for farmers in just seven countries could easily exceed $1.2 billion per year (3 times the amount spent on public agricultural research in the green revolution centres of the CGIAR or about half the yearly Canadian aid budget)." Finally, an industry claim is presented and disputed: "Corporations are now arguing that Terminator technology could be used to stop unwanted contamination from genetically modified plants. This is not true. Not only will Terminator fail miserably as a biosafety tool, it poses its own biosafety risks."

The organizations involved are a melange of interests from farming to food safety, the environment, and human rights, finding common cause in this technology and adept at using a broad range of media technologies to communicate, disseminate, and mobilize. Ongoing monitoring also occurs through regular communiqués, with updates on research and other market or policy occurrences. For example, the ETC Group's report *Terminator: The Sequel* (ETC Group 2005) describes the crop of genetic engineering technologies that are more recently being promoted as a biosafety solution to the spread of transgenes from GM crops, trees, and pharmaceutical-producing plants. The twenty-eight-page *Communiqué* begins with an examination of the European Union's Transcontainer project aimed at developing GM crops and trees that could be biologically contained through "reversible transgenic sterility," represented as the next generation or "sequel" version (ETC Group 2005). The EU's three-year Transcontainer project, part of the European Union's Sixth Framework Programme, supports the coexistence of GM crops and non-GM crops. The ETC communiqué disputes the Transcontainer project claims that its aim is not to restrict seed use but to contain transgenes and that the technology under development differs from terminator technology because the seeds' sterility will be "reversible," so that seed fertility can be recovered, most likely through the application of a chemical (Transcontainer 2006-9). This is countered by the ETC Group, which offered the following assessment:

> A scenario in which farmers would have to pay for a chemical to restore seed viability creates a new perpetual monopoly for the seed industry. Even if these *'Zombie seeds'* are not being designed with the intent to restrict seed use, the reality is that farmers will end up having to pay for the privilege of restoring seed fertility every year. (ETC Group 2005)

The Transcontainer project scientists responded in turn to the ETC Group, arguing that research on these biological containment mechanisms was designed "to facilitate the coexistence of GM and non-GM crops in European agriculture" and disputing the allegation of "zombie seed" creation (Transcontainer 2006-9).

Several features of this case are relevant in describing the evolution of participation by stakeholder groups on this issue. First is the set of mechanisms through representational strategies in public arenas such as the media and other public forums. Tactics of naming ("suicide seeds"), framing ("a social justice issue"), and claims making ("an assault on poor farmers' rights") are critical to elaborating the contention that such technologies are risky, unjust, and unnecessary. The case illustrates the generation, diffusion, and utility of representations, including mobilizing and countermobilizing ideas and meanings. These strategies also provided an easy hook for the media, which played up battlefield scenarios, using images of ongoing "fierce skirmishes against terminator technology" (Greenwood 1999, 39) and referring to these battles as the "seeds of discord" (Service 1998, 850). An example of an important opportunity aperture was the growing international interest in biodiversity, formalized through the Convention on Biological Diversity, an important environmental context that allowed for the recognition of the need to protect genetic resources.

Second is the relational dimension with the successful enrolment of key institutional players, including CGIAR, the FAO, and the Rockefeller Foundation. Such validation by powerful outsiders was critical to maintaining the ban on these technologies, providing important "certification" of the claims and their claimants. Another relational strategy is the linkage with numerous other like-minded organizations. Use of the Internet allows a large and diverse number of NGOs that have found common cause on the terminator issue to maintain and keep track of the ongoing storyline, shared continuously with other members of the network and ensuring constant attention, interest, and vigilance. Campaigns are in English, Spanish, French, and Portuguese to ensure world-wide accessibility.

Third is the alternative discourses that challenge supportive technological frames and, in turn, promote further international discussion and debate on these technologies. It is unclear how long this moratorium will last. However, when logics or claims are incommensurate, opportunities for change – whether in policy (Armstrong and Bernstein 2008) or technological design – can arise.

Case 2: Patient Organizations and Rare Diseases – Reframing Risk, Reproducing Knowledge

Activities of grassroots stakeholder organizations in the human genetics arena illustrate another aspect of the dynamics of public participation that focuses on the entire innovation trajectory, from knowledge production to knowledge translation. Knowledge production and translation comprise a constellation of ideas, resources, standards, and procedures; legislative and policy frameworks; and cultural climates, and patient organizations have been most instrumental in reframing knowledge production processes with this broader view. Organizations such as PXE International (Terry et al. 2007), the Chromosome 18 Registry and Research Society (Cody 2006), and similar groups focused on rare diseases, a large majority of which are genetics based, are reflective of stakeholders transforming the way in which knowledge is produced and translated in health systems.

The story of PXE International reflects the emergence and transformation of one patient organization and its connections with a large network of similar patient organizations. PXE International was founded by patients diagnosed with pseudoxanthoma elasticum or PXE, a disease that causes central vision loss, subsequent blindness, gastrointestinal and cardiovascular disease, and other manifestations. Like other individuals suffering from rare diseases, these patients faced a number of challenges, including minimal or conflicting medical advice and limited funding and pools of participants for research, resulting in little interest from the research community (Terry and Terry 2006; Terry et al. 2007; Wästfelt, Fadeel, and Henter 2006). The creation of PXE International focused on developing a blood and tissue bank that included a registry of well-annotated samples and whose parameters were defined by the patient community. Collaborations with the research community, the provision of research funding, and access to tissue samples through contractual arrangements, including material transfer agreements (Terry and Terry 2006), subsequently led to the discovery of the gene associated with PXE. This gene was co-patented by PXE International and its scientific collaborators, marking the first time that a patent was held by a patient organization. This organization is also working with the FDA to approve a genetic test for PXE, another first in terms of a stakeholder, non-profit, patient organization applying for FDA's diagnostic review and device clearance.

The larger and longer-standing coalition of genetics advocacy organizations under the umbrella of US-based Genetic Alliance, a coalition of several hundred rare genetic disease organizations and the European Organization for Rare Diseases (EURORDIS), also a patient-driven alliance of organizations covering over 1,000 rare diseases, have developed partnerships with scientists, regulatory authorities, pharmaceutical companies, and small and large biotechnology enterprises that have networked to raise pools of funding and research resources through tissue banks for basic research and following through to drug development (Terry et al. 2007; Wästfelt, Fadeel, and Henter 2006).

These organizations have significantly revamped research procedures, including rewriting standardized protocols for informed consent and creating formal agreements such as memoranda of understanding and material transfer agreements and licensing procedures to govern the use of tissue samples. What these patient networks have called "disruptive innovation" (Terry et al. 2007) characterizes the alternative ways in which knowledge production and translation are being carried out. Nowhere in the research community has such a full web of knowledge, work, and players been instituted that incorporates the research process from knowledge production to clinical use. This is evident in practices such as patients or family members patenting knowledge and driving research directions or group members participating as authors and co-authors in scientific publications and partnering with industry to follow through with developing diagnostic tools and therapies (Einsiedel 2009; Novas 2006).

In their pursuit of a new model of doing research and creating knowledge, these organizations promote alternative contentions of risks, rights, and responsibilities: "That perspective now includes a willingness to accept more and different kinds of risks, a willingness to sacrifice more to attain benefits, and understanding our shared inheritance in a profound way" (Terry and Terry 2006, 413). Such a view, in turn, fosters different governance arrangements for scientific research and ways of producing knowledge.

How do the mechanisms of contention apply in the case of patient groups concerned about rare genetic diseases? Their processes began with the cognitive redefinition of rare diseases as an opportunity rather than a condition of neglect and disapprobation. Families affected by accidents of genetic mutations organized in their common cause, initially around their affliction of interest. A shift in scale occurred through the umbrella of Genetic Alliance and the realization that different rare disease groups had more interests in common, a model that in turn has inspired similar models across the Atlantic. Scientific advances in genomics, from the sequencing of the human genome to technological advances in tissue banking, have provided the environmental triggers that have contributed to an "opportunity spiral" of consequences. Their establishment of relationships with researchers and their development of research networks, tools, and institutional arrangements to facilitate innovation from knowledge production to translation further reflect the relational features of mobilization.

Discussion

I began this chapter by suggesting that our concepts of publics and participation be expanded if we are to more fully understand trajectories of technology development. The examination included organized publics or individuals or groups who organized themselves around a common interest. In posing the question of how trajectories of science and technology take place, we can understand the activities of such groups as ways of participating in the shaping of such trajectories. Technology development can be an arena of contentious politics, and our understanding of participation can only be enhanced with a deeper exploration of the mechanisms beneath the processes of participation. The framework offered by the dynamics of contention has allowed me to elaborate on what happens when organized groups or groups of people who organize themselves around a social problem take part in contentious politics. Whether the groups end up stopping a technology in its tracks or appropriating the reorganization of technologies to advance their interests is less important in this context than understanding mechanisms of participation in public arenas.

I chose two illustrative cases to suggest that contentions about technologies in public arenas beyond the formal ones, arenas that can include supranational spaces or arenas not typically associated with participatory projects (labs, patent offices, journals), can reveal more possibilities for participation in technology development and assessment. Events in the broader environment can bring about the initial recognition of an opportunity to extend previous mobilization efforts such as the scientific development of gene use restriction technologies. For rare disease patient groups, it can be the emotional trigger of a child born with a genetic condition. In both instances, an opportunity spiral can happen.

How these opportunities are translated into further openings for growth and expanded spheres of activity – whether defined by increased memberships, fundraising, extensions of sociability through larger networks, certification by powerful institutions, or the extension and adoption of claims and frames – in turn help to elucidate and redefine the potential impacts of such participation on emerging technologies in the public sphere.

REFERENCES

Aoki, K., and K. Luvai. 2007. "Seed Wars: Controversies over Access to and Control of Plant Genetic Resources." In *Patents and Trade Secrets.* Vol. 2 of *Intellectual Property and Information Wealth Issues and Practices in the Digital Age,* edited by P. Yu, 334-85. Westport, CT: Praeger.

Armstrong, E.A., and M. Bernstein. 2008. "Culture, Power, and Institutions: A Multi-Institutional Politics Approach to Social Movements." *Sociological Theory* 26, 1: 74-99.

Ban Terminator. 2007. Endorsements and Issues http://www.banterminator.org/.

Blume, S. 1997. "The Rhetoric and Counter-Rhetoric of a 'Bionic' Technology." *Science, Technology, and Human Values* 22: 31-56.

Bueckert, D. 2006. "Ottawa Seeks 'Suicide Seeds' Testing." *Globe and Mail,* 21 March, A12. http://www.theglobeandmail.com/.

Buttel, F.H. 1997. "The Global Politics of GEOs." In *Engineering Trouble,* edited by R.A. Schurman and D.D. Kelso, 152-65. Berkeley: University of California Press.

CGIAR. 1998. *CGIAR Policy Statement on Genetic Use Restriction Technologies.* Booklet of CGIAR Centre Policy Instruments, Guidelines, and Statements on Genetic Resources, Biotechnology, and Intellectual Property Rights, System-Wide Genetic Resources Programme (SGRP). Rome: CGIAR.

Cody, J. 2006. *Who We Are.* http://www.chromosome18.org/.

Crucible Group. 1994. *People, Plants, and Patents: The Impact of IP on Biodiversity, Conservation, Trade, and Rural Society.* Ottawa: IDRC.

Dewey, J. 1927. *The Public and Its Problems.* Chicago: Swallow Press.

Einsiedel, E.F. 2009. "Stakeholders and Representation." In *Handbook of Genetics and Society: Mapping the New Genomic Era,* edited by P. Atkinson, P. Glasner, and M. Lock, 187-202. London: Routledge.

Einsiedel, E.F., M. Jones, and M.J. Brierley. 2011. "Cultures, Contexts, and Commitments: Public Participation and Xenotransplantation Policy in the U.S., U.K., and Canada." *Science and Public Policy* 38, 8, 619-28.

ETC Group. 2005. *Terminator: The Sequel.* http://www.etcgroup.org/.

–. 2006. "Monsanto Announces Takeover of Delta and Pine Land and Terminator Seed Technology (Again)." News release, 16 August.

FAO. 2001. *Report of the Panel of Eminent Experts on Ethics in Food and Agriculture.* http://www.fao.org/.

Fung, A. 2003. "Recipes for Public Spheres: Eight Institutional Design Choices and Their Consequences." *Journal of Political Philosophy* 11: 338-67.

Funtowicz, S., and J. Ravetz. 1993. "Science for the Post-Normal Age." *Futures* 25, 7: 739-55.

Gibbons, M., et al. 1994. *The New Production of Knowledge: The Dynamics of Science and Research in Contemporary Societies.* London: Sage.

Greenwood, J. 1999. "Terminator Gene Trips Alarm." *National Post,* 20 February, A14.

Habermas, J. 1989. *The Structural Transformation of the Public Sphere.* Cambridge, UK: Polity Press.

Hendriks, C. 2006. "When the Forum Meets Interest Politics: Strategic Uses of Public Deliberation." *Politics and Society* 34, 4: 571-601.

Jefferson, R., D. Blyth, C. Correa, G. Otero, and C. Qualset. 1999. "Genetic Use Restriction Technologies: Technical Assessment of the Set of New Technologies which Sterilize or Reduce the Agronomic Value of Second Generation Seed, as Exemplified by U.S. Patent No. 5,723,765, and WO 94/03619." Expert paper prepared for the Secretariat, 30 April. UNEP/CBD/SBSTTA/4/9/Rev.1.

McAdam, D., S. Tarrow, and C. Tilly, 2001. *Dynamics of Contention.* Cambridge: Cambridge University Press.

Michael, M. 2009. "Publics Performing Publics: Of PIGs, PIPs, and Politics." *Public Understanding of Science* 18, 5: 617-31.

Novas, C. 2006. "The Political Economy of Hope: Patients' Organizations, Science, and Biovalue." *Biosocieties* 1: 289-305.

Oliver, M.J., J.E. Quisenberry, N.L.G. Trolinder, and D.L. Keim. 1998. "US5723765: Control of Plant Gene Expression." *USPTO Patent Full Text and Image Database.* http://patft.uspto.gov/.

Pinch, T., and W. Bijker. 1987. "The Social Construction of Facts and Artifacts: Or How the Sociology of Science and the Sociology of Technology Might Benefit Each Other." In *The Social Construction of Technological Systems: New Directions in the Sociology and History of Technology*, edited by W. Bijker, T. Highes, and T. Pinch, 17-50. Cambridge, MA: MIT Press.

RAFI. 1999. "RAFI's Impact – 1999." Insert to RAFI's *1998-1999 Annual Report*. http://www.etcgroup.org/.

Reuters. 2000, 8 February. "GMOs Could Help War on Hunger: Interview with FAO." In *Terminator Two Years Later: RAFI Updates on Terminator/Traitor Technology*. Report prepared in preparation for the Fifth Conference of Parties to the Convention on Biological Diversity. Nairobi, Kenya, 22.

Rowe, G., and L. Frewer. 2005. "A Typology of Public Engagement Mechanisms." *Science, Technology, and Human Values* 30, 2: 251-90.

Service. 1998. "Terminator Seeds Sow Discord." *Science*, 27 October, 850.

Steinbrecher, R., and P.R. Mooney. 1998. "Terminator Technology: The Threat to World Food Security." *Ecologist*, 1 September, 275-84.

Tarrow, S., and D. McAdam. 2004. "Scale Shift in Transnational Contention." In *Transnational Protest and Global Activism*, edited by D. Della Porta and S.G. Tarrow, 121-35. Lanham, MD: Rowman and Littlefield.

Terry, S.F., and P.F. Terry. 2006. "A Consumer Perspective on Forensic DNA Banking." *Journal of Law, Medicine, and Ethics* 34, 4: 408-14.

Terry, S.F., P.F. Terry, K. Rauen, J. Uitto, and L. Bercovitch. 2007. "Advocacy Groups as Research Organizations: The PXE International Example." *Nature Reviews Genetics* 8: 157-64.

Tilly, C., and S. Tarrow. 2006. *Contentious Politics*. Boulder, CO: Paradigm.

Transcontainer. 2006-9. http://www.transcontainer.wur/.

Wästfelt, M., B. Fadeel, and J. Henter. 2006. "A Journey of Hope: Lessons Learned from Studies on Rare Diseases and Orphan Drugs." *Journal of Internal Medicine* 260: 1-10.

17

Public Engagement with Human Genetics
A Social Movement Approach

ALEXANDRA PLOWS

Public engagement with science is a key discourse in academia, science/ industry, and especially policy (Bucchi 2004; Leshner 2003; Rowe et al. 2005; Wynne 2006). But what is meant by such "public engagement," and what is its seen purpose? Informed by theories of deliberative democracy (Dryzek 2000; Fischer 2000; Habermas 1987) and by social movement theory (Diani 1992; Melucci 1996; Tarrow 1998), this chapter argues that a distinction can be made between public engagement as a specific set of *policy practices* (governance) and public engagement as a *social movement* in which engagement is understood as plural forms of mobilization through which publics "frame" issues on their own terms, generally but not exclusively outside the policy sphere. Public engagement with human genetics as a social movement is thus defined here as grassroots, self-initiated forms of mobilization, including protest activity. Public engagement is generally understood in policy and many academic accounts as a specific policy tool (Martin 2009) whereby policy makers seek to engage the public through (for example) consultation processes that have specific and fixed aims and agendas; public engagement as a social movement is a more active, grassroots-led form of engaging publics. Public engagement as a policy practice is, therefore, usually seen as a means to an end, whereas social movement public engagement can be an end in itself. These are citizens engaging in deliberative democracy on their own terms, contributing to citizenship stakes both through the forms of participation that they enrol and the issues that they raise through this process. Such engagement performs important functions, such as identifying important ethical and other concerns, civil society capacity building, challenging power relations and cultural assumptions ("challenging codes" [Melucci 1996]), and creating social change.

Embedded within the question of what public engagement is, and is for, is the issue of who "the public" are. Academics have differentiated between the general or lay public and specific publics as expert, engaged stakeholders (Collins and Evans 2002; Evans and Plows 2007; Fischer 2000; Wynne 1996, 2006) and/or interest groups (Tait 2001). Qualitative research conducted during 2003-7 (*Emerging Politics of New Genetics Technologies*),[1] which forms the evidence base for this chapter, traced the social dynamics of multiple publics as they engaged with human genetics, identifying core areas of public interest and concern. This research identified that there are multiple publics, multiple issues at stake, many entry points into complex debates on human genetics, and many ways of framing them. Importantly, not just interest groups but also many different sorts of publics are predisposed to engage with human genetics if personally motivated by specific issue triggers or by how these issues emerge and are framed as public debates or emerge as points of consumption. Such lay publics can develop "organic expertise" (Evans and Plows 2007). For example, a member of the public asked to donate DNA to the Wellcome Trust biobank could then become an engaged, expert stakeholder. Someone buying a genetic test kit online can also be described as an engaged member of the public.

A social movement has been defined as "a network of informal interactions between a plurality of individuals, groups and/or organisations, engaged in a political or cultural conflict, on the basis of a shared collective identity" (Diani 1992, 13). Although relatively little current public engagement with human genetics can be defined strictly as a social movement,[2] the melange of social network interactions, occasional protest events, and other forms of mobilization that different publics are catalyzing and participating in are all types of social movement. Social movement theory has provided a useful set of tools to understand many different types of social mobilization (Doherty et al. 2003). Identifying and analyzing public engagement outside the policy sphere is sociologically important in terms of identifying other engaged publics, their mobilization repertoires (Tarrow 1998), and their reasons for mobilizing, including issues thought to have been framed out of policy processes (Plows 2007, 2010). Studying social movement forms of public engagement clearly can inform policy and political directions and debates more generally about science and society. For analysts and policy makers wishing to understand which issues are at stake for the public, studying such activity thus provides an opportunity to broaden the stakes of the debate and incorporate the views of other stakeholders into the decision-making process.

Some groups and networks deliberately choose to mobilize outside the policy arena in relation to human genetics, as other public groups do in relation to many other issues and situations, such as environmental controversies (Barnes, Newman, and Sullivan 2007; Doherty et al. 2003; Doherty, Plows, and Wall 2003; Fischer 2000). This type of public engagement as social movement is often representative of a reflexive critique of power relations and agenda setting in specific contexts, such as the environmental and health risks of nanotechnology or concerns about prenatal genetic screening policy. This chapter provides a narrative of a specific case study cluster of emergent social networks, groups, and individuals that engaged in debates about human genetics but whose issue frames were, broadly speaking, less well represented in consultation processes and other policy/public settings. Examples of protest activity and other public engagement outside the official policy process are provided, identifying "prime movers" (McAdam 1986) and "early risers" (Tarrow 1998), who set the stakes of the debate on their own terms, for example through protests, publications, and workshops. Importantly, these repertoires of action are not restricted to groups who mobilize in opposition to different human genetics applications (the focus of this chapter) but include groups who mobilize to support them, such as patient groups, parents of sick children, and so on (Evans and Plows 2007; Plows 2010). Such events shed light on the more latent and hidden nature of informal networks of activist and campaign groups and show how activists develop capacity.

Public Engagement as Policy Practice

Many academic commentators have noted that public engagement as policy practice performs a set of specific functions that have important benefits, including addressing an identified "democratic deficit" (Abelson et al. 2003; Dryzek 2000; Pratchett 1999) through increasing public participation in governance and policy processes, with a potential increase in the accountability of governance. There are acknowledged limitations, however, in terms of the efficacy of the process and what it is actually set up to deliver in terms of policy outcomes (Martin 2009), leading to discrepancies (and occasional accruing conflicts) between what publics expect from such processes and what politicians and policy makers expect from them. Irwin (2007) argues that there has simply been a shift to addressing a public "trust deficit." Thus, while public engagement has become part of a new high-tech governance, the imperatives governing technology remain unchanged. There are also concerns about how publics and their potential inputs are framed. These issues have implications for citizenship stakes in relation to the impact of public engagement – what is debated, at what point in the process, by whom, and with what outcomes. This is particularly important given that definitions of public (citizen) duties and responsibilities, as well as rights, are at stake in the field of genetics, health, and medicine; for example, it has been suggested that we have a "duty" to participate in biomedical research (Harris 2005). Thus, though some commentators are optimistic about the role of "bio citizenship" (Rose and Novas 2004), such discussions often assume that participation in deliberative democracy is a level playing field; this is, to say the least, highly questionable (Plows and Boddington 2006).

Thus, such public engagement is part of the democratic governance of technologies, aiming to enrol identified stakeholders in set processes. It is a specific practice with specific aims and agendas. For example, a 2007 UK government report, based on findings from a public engagement process on the topic of public consultation, stated that "consultation is ... a form of engagement that is appropriate when the policy process is already underway and there is an intention to make changes or deliver specific outcomes. It therefore does not invite an open debate on very broad areas of public policy, nor does it empower those who participate with the final decision" (Department for Business Enterprise and Regulatory Reform (BERR 2007, 7). This report on public consultation identified significant "consultation ennui" among stakeholders because the limits of the process inevitably led to a failure to respond to issues raised and changed aims and objectives accordingly.

Importantly, academic and stakeholder calls for better public engagement posit the "deficit model" of public understanding of science (Wynne 1996) and, drawing on concepts of patient "embodied expertise" (Kerr et al. 1998) have led to more participatory models of public engagement being developed to encourage the involvement of expert publics and engaged publics in the regulatory process (Epstein 1995; Evans 2008; Evans and Plows 2007; Rogers-Hayden and Pidgeon 2006; Singh 2008). Bucchi (2004) has identified and recommended the development of further cross-talk among publics and scientists to facilitate knowledge transfer. A related development has been a discourse on "upstream" public engagement (Ferretti and Pavone 2009; Wilsdon and Willis 2004; Wynne 2006), which explicitly seeks to enrol stakeholders early on in the regulatory process. However, for some, this has significant limits:

> Demos has ... produced a report, *See through Science*, that calls on industry, government and scientists to involve concerned groups in shaping research into problematic subjects such as nanotech at a much earlier stage than commercialisation. The problem with this argument is that ... development ... seen as "concerning" ... usually happens well after industry and government have set their targets. (Corporate Watch 2005)[3]

Thus, this is still "downstream" engagement and clearly relates back to the limited remit of any accruing consultation process – the "end of pipe" nature of public engagement as policy practice, which only occurs once a specific trajectory of technological development is well in process. There is a linked set of public concerns over power relations and agenda setting; who and what are really upstream?

Policy-led public engagement with human genetics thus has a specific modus operandi; it operates within a specific political landscape, with a specific remit, which shapes the ensuing public response. In various fields such as public health, other academics have also commented on the limits of outcomes (impacts) of such public consultation and the disjuncture between rhetoric and reality (Barnes, Newman, and Sullivan 2007; Hodge 2005; Martin 2009; Williams 2004). Such public engagement, by its very nature, is likely to miss, and indeed to frame out, much of what is being said at the grassroots and to have a limited impact on decision making, though this varies from context to context, and some stakeholders have identifiably better access, and greater legitimacy, than others (Plows 2010). Many activists and campaigners in the case study discussed here are suspicious of being enrolled in policy processes that they see as merely furthering governance through the acquisition of legitimacy because of claims to have involved the public in decision making. They either engage in such a process reluctantly or, in many cases, are openly hostile and not only withdraw from policy-led forms of "participatory democracy" but can also mobilize in direct opposition to the state and market, for example targeting specific corporations (Chesters and Welsh 2006).

Public Engagement as Social Movement

The remainder of the chapter provides a number of examples of public engagement as social movement, narratives that outline the motivations of participants themselves, showing that publics who engage outside the policy process do so for a number of reasons, such as "tilting the frame" (Steinberg 1998) – seeking to reframe the debates on their own terms, not those set by powerful elites. Studying the nature of such engagement contributes to an understanding of how people negotiate highly complex and ambivalent terrains. Although participants usually want to achieve specific outcomes, such activity also has intrinsic value as part of social life that has meaning for participants and helps them to develop capacity, such as developing knowledge stakes and network contacts.

The case study discussed here involved groups, networks, individuals, and organizations with fluid or "weak ties" to each other (Granovetter 1973). They can broadly be defined as approaching human genetics from an environmental and social justice perspective, often explicitly supportive of disability rights, in that issues of health and identity raised by human genetics applications are always situated in a framework that takes account of the social environment.

> There's such a fixation these days on genomics, and the gene has all the answers. Whereas the glaringly obvious health issues, like ... malnutrition ... in moms and kids, ... are blatantly ignored. ("Alice," bio-informatician)

> My life in the last two years has been 100 percent better because I've been in housing which meets my needs ... They're spending huge wads of money on [genetics research], when they could be fixing the day-to-day problems of people that have all sorts of conditions by spending that money elsewhere. ("Sally," disability rights campaigner)[4]

For many of these groups, networks, and individuals, the governance of human genetics and technoscience generally, including agenda setting and how the issue is framed by the policy process, is a key motivation for action, which is initially catalyzed by a specific issue. The following quotations (taken from interviews unless otherwise indicated) discuss campaigners' concerns about the framing of debates.

"Simon," a genetics watchdog campaigner, argues that "the debate is framed as a simple binary opposition ... It's painful to try and explain the nuances." "Lucy" notes how "the [stem cell] debate has been framed in a certain way, you're either for or against ... So other issues, other considerations, ... cannot be expressed." An international meeting of civil society NGOs asked these questions: "How do we avoid being cast as opponents of medical research and individual liberties? ... How much do we focus on the technologies themselves, and how much on the social justice and global equity values that motivate our concerns?" (text from civil society genetics event, Berlin, 2003). This quotation also emphasizes how people struggle to frame issues on their own terms, not least because of the complexity of the issues at stake.

Some campaigners do more than critique power relations – they try to redress the power imbalance through direct action. Activists in France in 2004 occupied a site earmarked for a major nanotechnology research site, stating that they did so as a means of "direct democracy," signalling a lack of trust in the democratic process: "Potential consequences and benefits drawn [sic] by research can't be debated or controlled by the populations [sic] ... Shutting this site ... is attempting to stop a project that we refuse ... It seems to us essential to interrupt 'development's' headlong·rush."[5]

Such direct democracy contributes to the debates on its own terms and is seen by participants and advocates as an important form of citizenship necessary to redress power imbalances within existing governance frameworks, which they argue push technologies along set pathways. "Mike," a technology watchdog campaigner, argues that "what we need to get to is ... the politics of new technologies, ... a real live politics of how new technologies impact on society, how society has some control over that." These are extremely important contributions to citizenship and science debates and address the "bigger picture" underlying different scientific controversies, positions that are also reflected in established Science and Technology Studies (STS) perspectives (Wynne 2006).

Case Study Examples
Repertoires of action common to many different public groups can be summarized as

- networking (informal and formal) and capacity building within groups and networks; for example, workshops, blogs, and e-networks;
- protests;
- lobbying (e.g., MPs, "big pharma");
- input into policy/regulatory processes (as expert witnesses, responding to different consultations, etc.);
- public education, awareness raising, and engagement work; putting on events, literature, websites (links to networking); and
- constructing, and interacting with, others as "allies or enemies" (sometimes this is counterintuitive and context dependent).

Importantly, many people as individuals or campaign groups also contribute to the policy process through consultation processes, for example, and are thus engaged in many "hybrid" forms of public engagement. Other groups, networks, and individuals deliberately stay outside the policy process and participate only through forms of engagement that they generate themselves.

During the course of the research project, a number of workshops and other events, including new media interaction (web chats, websites, etc.) organized by public groups and networks for their own benefit, were identified, and some were participated in. These events can be seen as "convergence spaces" (Routledge 2003), which both reflect the current state of network capacity and catalyze further mobilization. They tend to be consciously aimed at capacity building within existing networks and groups; they are put on by the organizations and networks foremost as an internal resource but also as a recruitment strategy. Sometimes these events have an articulated aim of generating direct action (protest) as an outcome. Depending on the circumstances, some of these events are only advertised internally, and knowing about them can often mean that one needs pre-existing knowledge of the particular social network, making these events difficult (and in some cases impossible) to access by "outsiders" and thus challenging sites to study. An example is a workshop on enhancement held in Oxford in 2006 that was attended by a number of prime movers in the UK social and environmental justice movement. This event was by invitation only and thus restricted to people within pre-existing networks in which friendship, trust, and a shared background of action and campaigning were key to their presence at the event.

The genetics and science workshops held during the European Social Forum (ESF) in London in 2004 are another example of public engagement for building movement capacity. The ESF is a roughly annual event consisting of European grassroots networks, groups, NGOs, and far-left political parties; it is an important event in the calendar of global civil society networks that have been termed the "anti" (alter) globalization movement. The ESF is a "convergence space" (Routledge 2003; Welsh, Evans, and Plows 2007) that aims to develop network capacity and generate action. The list below gives the titles and organizers of the workshops provided at the ESF on the topic of human genetics, bioscience more generally, and science and society. It shows which human genetics developments/applications were catalyzing mobilization at the time, how they were framed, and which groups and networks were prime movers and early risers shaping the debates. They show the process of meaning construction in action at an early stage, showing how networks and groups take the opportunity to build capacity through information sharing and discussion.

1 "Developments in Human Genetics" (GeneWatch UK, Human Genetics Alert, Institut Mensch Ethik und Wissenschaft, Gen Ethisches Network)
2 "Bar Coding People: Individualized Health Care or Money Making Scam?" (GeneWatch UK)
3 "Human Cloning and Genetic Engineering: What's at Stake?" (Human Genetics Alert)
4 "Prenatal Screening: Eugenics or Women's Rights?" (Human Genetics Alert)

Related workshops included "Resisting Corporate Monopolies and New Enclosures" (Action Group on Erosion, Technology, and Concentration [ETC Group], Green Party of Europe, Protimos). This was one of several workshops held specifically on nanotechnology and converging technologies. Science and society workshops included "What Research Policies Are Appropriate in Another Europe?," "Science and Citizenship," and "A European Science Social Forum (ESSF)." These three workshops, by far the most "official" with academic and relatively "big name" speakers and international NGO presence, had a definite policy orientation. They were the most strategic in terms of wanting to enter the policy arena and reframe the stakes of policy debates. Soon afterward, at a "Science in Society"[6] high-profile public participation event in Brussels, some of the same prime movers "launched" the ESSF, presenting a discussion document on Framework 7 (FP7), the European funding scheme, criticizing how budgets were allocated and identifying a number of concerns with the emphasis on funding science and technology projects.

Another important form of social movement is more "outward facing," awareness raising, and networking. This covers a wide range of different types of information creation and dissemination, including e-networks, blogs and websites, publications, public events, and specific campaigns. This cluster of activities tends to represent the more public-facing aspect of network, campaign group, and NGO activity, for example the publications produced by the UK NGO GeneWatch, the international NGO ETC Group, or (to give an example from a different set of civil society networks) the "pro-genetic" organization Genetic Interest Group. The production of these resources from networking to publications is aimed at raising awareness among the general public, increasing membership (whether formal or informal) of specific campaign groups and networks, and putting agendas and issues into the public sphere for further debate and/or to affect the political direction of events. Such activity is often catalyzed by specific policy developments, such as consultation by the Human Fertilisation and Embryology Authority (HFEA 2006) on the use of "donated" eggs for stem cell research or developments in the private sphere, such as the increase in commercially available genetic tests, which has catalyzed a number of reports by GeneWatch. Engagement of this type can be short term (linked to a timeline around a specific campaign over a policy decision), long term, or often a mix.

Protest events are another important form of public engagement as social movement. Protest activity often performs a symbolic function (Melucci 1996) and not always (or solely) a disruptive one. Looking at the form, target, and rationale for a protest event and the makeup of participants can inform an understanding of which issues are at stake. UK protest events about human genetics often were not identified during the research time frame. It is possible that relatively few protest events occurred because the stakes are still emergent and highly complex, and thus it is hard for affected or interested publics to know how and why to "draw lines" and focus on appropriate targets even where there is an identified grievance. Also, as the quotation from the Berlin civil society event given earlier points out, to voice opposition places one at risk of being seen as anti-technology, which many people found extremely problematic and a block on taking oppositional action. That issues are emergent and highly complex can also explain why several protest events were unusually hybrid in terms of the people who participated.

Three examples of protest events organized by the case study networks are given below. The first example is a protest organized by People against Eugenics (PAE) in October 2004. This was a small demonstration attended by genetics watchdogs, social justice and disability rights campaigners, and, significantly, some pro-life campaigners. The demonstration was a response to a conference held at the Royal Society entitled Ethics, Science, and Moral Philosophy of Assisted Human Reproduction, which had invited several speakers known to take controversial positions on issues such as the use of genetic reproductive technologies for screening out disabilities and the use of germline cloning for "human enhancement." The disability rights (DR) campaigners were particularly concerned that the conference had no one from a DR perspective speaking. Demonstrators handed out leaflets to participants from outside the premises in a low-key, good-natured event. An important finding was the presence of pro-life activists at this demonstration. They had been invited by some of the DR activists with whom they had had previous contact. Significantly, these pro-life activists were seemingly happy to hand out a leaflet, written by the PAE protest organizers, that clearly stated that those present were supportive of reproductive rights to choose – that is, they were not anti-abortion. The presence of the pro-life activists caused some consternation among the main organizers of the event, and there were several interesting conversations between pro-life activists and other campaigners about why being concerned about prenatal testing was more than simply an issue of "embryo rights." Campaigns that enrol issues important to a number of different "lifeworlds" (Habermas 1987) tend to attract "strange bedfellows" (Evans et al. 2006).[7] Such hybridity is prefigurative of, and further informs, an understanding of social complexity in this still emergent field.

A second example is a Down's syndrome protest that took place in 2003. The following account is taken from the press release of the protest, sparked by the then impending uptake of pre-implantation genetic diagnosis (PGD) as an "opt out" prenatal genetic screening program in the United Kingdom, and highlights (as the PAE action also showed) that the exclusion of certain perspectives from "official" and elite debates is itself a trigger for direct action. Like the PAE protest, it shows how people use protest events as a means of reframing debates that they believe have already been set, or from which they are excluded, or both.

> On May 19th [2003], a group of people with Down's Syndrome and their supporters disrupted the International Down's Syndrome Screening Conference at Regents College in London ... [telling] the doctors that [they] oppose[d] Down's Syndrome screening and that people with Down's Syndrome are people not medical problems ... The protesters ... had written to the conference organisers in advance and asked to speak, but were refused ... It is unacceptable that doctors discuss better ways of preventing people with Down's Syndrome being born, whilst excluding their voices from the debate. This runs directly counter to one of the main demands of disabled people: "Nothing about us without us" ... This should be the start of a national debate on prenatal screening.

Such action sent a strong signal about inclusion and involvement of stakeholders, and clearly it was a very important early warning about how to conduct debates and policy in sensitive and highly complex arenas. Significantly, the stakes raised by the participants and organizers had nothing to do with the status of the embryo (as most "ethical" debates about prenatal screening have been framed): they were about identity, value, and citizenship.

The third protest occurred over the use of unregulated nanoparticles. On 9 December 2004, THRONG (The Heavenly Righteous Opposed to Nanotech Greed) disrupted the Nanotechnology: Delivering Business Advantage conference in Buckinghamshire, presenting a "can of worms" award to one of the conference participants, Harry Swan, formerly of Monsanto. The symbolic and playful nature of much UK direct action (Szerszynski 2005; Wall 1999) was evident; the THRONG activists dressed as angels to disrupt the conference, stating that "where these nano fools rush in we angels fear to tread."[8] The identification of specific corporations' operations as a clear target for criticism and action, and the use of such action as a means of framing broader concerns about risk and uncertainty, echoed previous campaigns among the same networks focused on the operations of companies such as Bayer and Monsanto, with clear critical reference to processes of globalization. As the social movement literature would predict (Diani 1992; Doherty, Plows, and Wall 2003), activists with pre-existing social bonds developed through previous "cycles of contention" (Tarrow 1998), and new actors brought into these networks, are developing resources and strategies to oppose the development and commercialization of nanotechnology and converging technologies. In this respect, it is noteworthy that this protest took place two months after the ESF workshops. Although the NGOs tend to have a more formal and "static" identity, the more radical networks tend to develop multiple micro-identities and campaign "brand names." They reflect the more informal, often "biodegradable," processes of social network interaction that spawn specific direct action events.

Conclusion

Based on case study work among UK publics engaging with human genetics issues, this chapter has tackled what public engagement is and what it is for, differentiating between public engagement as policy practice and public engagement as social movement. Participants in policy-focused deliberative democracy processes enter a terrain that is not neutral; the technologies have already emerged along specific trajectories or are projected to, given the political and economic investments in them. Thus, public engagement as policy practice is inevitably subject to major power imbalances, so the extent to which publics can influence emergent technological trajectories through such engagement processes is questionable (Plows and Bodding-ton 2006; Rowe et al. 2005). Furthermore, many publics believe that the debates about these technologies have been framed in terms that are too narrow or not conducive to discussing issues that they believe to be at stake. Indeed, policy-led public engagement can frame out important perspectives, perhaps especially when applying outcome-focused approaches in complex and ambivalent terrain. Thus, many are questioning the impact that such engagement is having and what its purpose is – a topic not confined to public engagement with genetics or even science but relevant to many other fields where questions about the relationship, and the nature of the contract, between the state and society are raised by the use of public engagement by the state (Dryzek 2000; Fischer 2000; Martin 2009; Williams 2004; Wynne 2006).[9]

Frustration with state-led deliberative democracy processes is the reason that some public groups are deliberately choosing to mobilize outside the policy arena. Public engagement as social movement is often representative of a reflexive critique of power relations and agenda setting and constitutes an important contribution to citizenship stakes simply by making this point about access to and impact on decision making and by bringing other perspectives and knowledge bases to the table. The case study groups and networks mentioned in this chapter are reflexively "tilting the frame" to reframe the debates on human genetics – which often tend to invoke broader debates about health and identity and include issues of social and environmental justice (Plows 2010). Public engagement defined as social movement can be an end in itself for those who are initiating or participating in these forms of action, and the above examples show that building capacity within their own networks and contributing to public awareness are important goals for many campaigners engaging outside the policy process. Clearly, many are using direct action as a means of directly influencing those with power and often (but not always) hope that this type of action will affect specific policy outcomes, for example by influencing public opinion, which in turn has impacts on policy.

Understanding social life more generally as public engagement can inform policy agendas and identify emergent public trends that are sociologically important in their own right. Public engagement as policy practice can thus be complemented and informed by other forms of data collection and sociological research. Ethnography is particularly suited to "messy" (Law 2006) social fields, identifying issue frames and prime movers as they emerge and engage. It can help to answer important "baseline" questions. What are the emergent issues for the public? Which publics are mobilizing and why? What are the key stakes that they raise? This is not a definitive methodology – this chapter represents a small soil sample of a complex terrain – but such techniques can be used to collate data that together with other qualitative and quantitative approaches can inform a more nuanced and sensitive debate on the issues and values at stake. However, better deliberative democracy models are no solution to the issue of how uneven power relations impact on the final decision-making process or indeed on how genetic technologies initially emerge as issues in the public arena. For this reason, publics who mobilize outside policy processes in opposition to the state and market perform an important form of citizenship – actively seeking to redress such power imbalances.

NOTES

1 See http://www.lancs.ac.uk/.
2 For further discussion on social movements, collective identity, and typologies in relation to public engagement with human genetics/bioscience, see Plows (2010).
3 See http://www.corporatewatch.org.uk/.
4 For a detailed discussion of the views and values of this case study cluster, see Plows (2010).
5 "Activists in Grenoble France Occupy Construction Site for Minatec Nanotech Centre 2004-12-13)," press release, http://www.indymedia.org.uk/.
6 See http://ec.europa.eu/.
7 Pro-life activists also set up the Hands Off Our Ovaries! campaign group, together with some feminists, about the use of women's eggs for research (Plows 2008).
8 From THRONG press release, quotation from THRONG "angel" Sarah Phimms. See http://www.angelsagainstnanotech.blogspot.com.
9 The debates in June 2010 about the rationale and remit for UK government public consultation over service cuts are a pertinent current example; see "George Osborne to Ask Public which Services Should Be Cut," http://www.guardian.co.uk/.

REFERENCES

Abelson, J., et al. 2003. "Deliberations about Deliberative Methods: Issues in the Design and Evaluation of Public Participation Processes." *Social Science and Medicine* 57: 239-51.

Barnes, M., J. Newman, and H. Sullivan. 2007. *Power, Participation, and Political Renewal: Case Studies in Public Participation*. Bristol: Policy Press.

BERR (Department for Business Enterprise and Regulatory Reform). 2007. *Effective Consultation: Asking the Right Questions, Asking the Right People, Listening to the Answers*. http://www.bis.gov.uk/.

Bucchi, M. 2004. "Can Genetics Help Us Rethink Communication? Public Communication of Science as a 'Double Helix.'" *New Genetics and Society* 23, 3: 269-83.

Chesters, G., and I. Welsh. 2006. *Complexity and Social Movements: Protest at the Edge of Chaos.* London: Routledge.

Collins, H.M., and R. Evans. 2002. "The Third Wave of Science Studies: Studies of Expertise and Experience." *Social Studies of Science* 32, 2: 235-96.

Corporate Watch. 2005. "Nanotechnology: What It Is and How Corporations Are Using It." *Corporate Technologies* 1. http://www.corporatewatch.org/.

Diani, M. 1992. "Analysing Social Movement Networks." In *Studying Collective Action,* edited by M. Diani and R. Eyerman, 107-35. London: Sage.

Doherty, B., M. Paterson, A. Plows, and D. Wall. 2003. "Explaining the Fuel Protests." *British Journal of Politics and International Relations* 5, 1: 1-23.

Doherty, B., A. Plows, and D. Wall. 2003. "The Preferred Way of Doing Things: The British Direct Action Movement." *Parliamentary Affairs* 56: 669-86.

Dryzek, J.S. 2000. *Deliberative Democracy and Beyond: Liberals, Critics, Contestations.* Oxford: Oxford University Press.

Epstein, S. 1995. "The Construction of Lay Expertise: AIDS Activism and the Forging of Credibility in the Reform of Clinical Trials." *Science, Technology, and Human Values* 20: 408-37.

Evans, R. 2008. "The Sociology of Expertise: The Distribution of Social Fluency." *Sociology Compass* 2: 281-98.

Evans, R., et al. 2006. "Towards an Anatomy of Public Engagement with Medical Genetics: Strange Bedfellows and Usual Suspects." In *New Genetics, New Identities,* edited by P. Atkinson et al., 139-57 London: Routledge.

Evans, R., and A. Plows. 2007. "Listening without Prejudice? Re-Discovering the Value of the Disinterested Citizen." *Social Studies of Science* 37, 6: 827-53.

Ferretti, M.P., and V. Pavone. 2009. "What Do Civil Society Organisations Expect from Participation in Science? Lessons from Germany and Spain on the Issue of GMOs." *Science and Public Policy* 36, 4: 287-99.

Fischer, F. 2000. *Citizens, Experts, and the Environment: The Politics of Local Knowledge.* Durham: Duke University Press.

Granovetter, M. 1973. "The Strength of Weak Ties." *American Journal of Sociology* 78, 6: 1360-80.

Habermas, J. 1987. "The Theory of Communicative Action: A Critique of Functionalist Reason." Vol. 2: *Lifeworld and System.* London: Polity Press.

Harris, J. 2005. "Scientific Research Is a Moral Duty." *Journal of Medical Ethics* 31: 242-48.

Hodge, S. 2005. "Participation, Discourse, and Power: A Case Study in Service User Involvement." *Critical Social Policy* 25: 164-79.

Human Fertilisation and Embryology Authority (HFEA). 2006. *Donating Eggs for Research: Safeguarding Donors.* http://www.hfea.gov.uk/.

Irwin, A. 2006. "The Politics of Talk: Coming to Terms with the 'New' Scientific Governance." *Social Studies of Science* 36, 2: 299-320.

Kerr, A., S. Cunningham-Burley, and A. Amos. 1998. "The New Genetics and Health: Mobilizing Lay Expertise." *Public Understanding of Science* 7: 41-60.

Law, J. 2006. *After Method: Mess in Social Science Research*. London: Routledge.

Leshner, A.I. 2003. "Editorial: Public Engagement with Science." *Science*, 14 February, 977.

Martin, G. 2009. "Public and User Participation in Public Service Delivery: Tensions in Policy and Practice." *Sociology Compass* 3, 2: 310-26.

McAdam, D. 1986. "Recruitment to High-Risk Activism: The Case of Freedom Summer." *American Journal of Sociology* 92: 64-90.

Melucci, A. 1996. *Challenging Codes: Collective Action in the Information Age*. Cambridge, UK: Cambridge University Press.

Plows, A. 2007. "You've Been Framed! Why Publics Mistrust the Policy Process." *Genomics Network Newsletter* 6: 22-23.

–. 2008. "Egg Donation in the UK: Tracing Emergent Networks of Feminist Engagement in Relation to HFEA Policy Shifts in 2006." In *Women in Biotechnology: Creating Interfaces*, edited by F. Molfino and F. Zucco, 199-219. New York: Scrivener Press.

–. 2010. *Debating Human Genetics: Contemporary Issues in Public Policy and Ethics*. London: Routledge.

Plows, A., and P. Boddington. 2006. "Troubles with Biocitizenship?" *Genetics, Society, and Policy* 2, 3: 115-35.

Pratchett, L. 1999. "New Fashions in Public Participation: Towards Greater Democracy?" *Parliamentary Affairs* 52: 616-33.

Rogers-Hayden, T., and N. Pidgeon. 2006. "Reflecting upon the UK's Citizens' Jury on Nanotechnologies: NanoJury UK." *Nanotechnology, Law and Business* 3, 2: 167-80.

Rose, N., and C. Novas. 2004. "Biological Citizenship." In *Global Assemblages: Technology, Politics, and Ethics as Anthropological Problems*, edited by A. Ong and S. Collier, 439-63. Malden: Blackwell.

Routledge, P. 2003. "Convergence Space: Process Geographies of Grassroots Globalization Networks." *Transactions of the Institute of British Geographers* 28, 3: 333-49.

Rowe, G., T. Horlick-Jones, J. Walls, and N. Pidgeon. 2005. "Difficulties in Evaluating Public Engagement Initiatives: Reflections on an Evaluation of the UK GM Nation? Public Debate about Transgenic Crops." *Public Understanding of Science* 14, 4: 331-52.

Singh, J. 2008. "The UK Nanojury as 'Upstream' Public Engagement." *Participatory Learning and Action* 58, 1: 27-32.

Steinberg, M. 1998. "Tilting the Frame: Considerations on Collective Action Framing from a Discursive Turn." *Theory and Society* 27, 6: 845-72.

Szerszynski, B. 2005. "Beating the Unbound: Political Theatre in the Laboratory without Walls." In *Performing Nature: Explorations in Ecology and the Arts*, edited by G. Giannachi and N. Stewart, 181-97. Frankfurt: Peter Lang.

Tait, J. 2001. "More Faust than Frankenstein: The European Debate about the Precautionary Principle and Risk Regulation for Genetically Modified Crops." *Journal of Risk Research* 4, 2: 175-89.

Tarrow, S. 1998. *Power in Movement: Social Movements, Collective Action, and Politics.* Cambridge, UK: Cambridge University Press.

Wall, D. 1999. *Earth First! and the Anti-Roads Movement: Radical Environmentalism and Comparative Social Movements.* London: Routledge.

Welsh, I., R. Evans, and A. Plows. 2007. "Human Rights and Genomics: Science, Genomics, and Social Movements at the 2004 London Social Forum." *New Genetics and Society* 26, 2: 123-35.

Williams, M. 2004. "Discursive Democracy and New Labour: Five Ways in Which Decision-Makers Manage Citizen Agendas in Public Participation Initiatives." *Sociological Research Online* 9. http://www.socresonline.org.uk/.

Wilsdon, J., and R. Willis. 2004. *See-Through Science: Why Public Engagement Needs to Move "Upstream."* London: Demos.

Wynne, B. 1996. "May the Sheep Safely Graze? A Reflexive View of the Expert-Lay Knowledge Divide." In *Risk, Environment, and Modernity: Towards a New Ecology,* edited by S. Lash, B. Szerszynski, and B. Wynne, 44-83. London: Sage.

–. 2006. "Public Engagement as a Means of Restoring Public Trust in Science: Hitting the Notes, but Missing the Music?" *Community Genetics* 9: 211-20.

18 The Limits of Liberal Values in the Moral Assessment of Genomic and Technological Innovation

CHLOË G.K. ATKINS

The history of liberalism and its values is an inspiring one – liberal ideals incited revolutions in the eighteenth century and have continued to instigate revolutions and civic reforms ever since. Canada, already a well-established democracy, adopted the Charter of Rights and Freedoms in 1982, creating a wave of socio-legal activism in which different people and minority groups have successfully pursued their rights in courts and tribunals. South Africa followed a similar path, establishing a post-Apartheid, liberal democratic constitution in 1996. Moreover, in the latter part of the twentieth century and the beginning of the twenty-first century, liberal values expanded beyond the politico-legal sphere and penetrated the moral decision-making ambit of individuals. This has been most obvious in the areas of health care and in scientific and social science research and innovation. Biomedical and research ethics committees have burgeoned in the past two to three decades. It is no longer possible to carry out any medical treatment and/or pursue any form of research without considering the rights of participants. Physicians, researchers, and scientists must either seek the "informed consent" of those involved or plan for and mitigate any potential harms for participants.

The discourse of human rights, equality, and individual autonomy has thus expanded into the realm of public and personal decision making, leading to increased efforts at public consultation in terms of emerging technologies, environmental degradation, and genomic innovation. In an egalitarian community, scholars as well as laypeople are now expected to have opinions about the introduction and incorporation of complex technology and the moral issues that arise from its use. As an academic focused on applied ethics, I have served (and continue to serve) on numerous ethics committees as part of this expansive application of liberal philosophy to moral-scientific problems. More personally, I have written about my own decision making when, as a family, we confronted cytogenic testing (Atkins 2008). More recently, I have been captivated by the ethical quandaries that we unwittingly find ourselves in on a daily basis.[1] But more importantly – and my focus in this chapter – I have become skeptical that liberal values facilitate moral decision making when they are applied to genomic and technological innovations. Paradoxically, I argue that liberalism actually encourages the pursuit of scientific and technocratic ends rather than the consideration of individual dignity – and liberal values, in this context, undermine moral considerations of respect for the individual rather than support them.

Given that this discussion begins where a previously published discussion leaves off, I will briefly summarize the main elements of that earlier piece. I am a member of a same-sex couple with four children, ages twenty-five, twenty-three, seven, and five. In 2002 and again in 2003, my spouse and I considered using prenatal testing during the pregnancies with our two youngest sons. As a couple, we were both well versed in moral philosophy as well as the medical procedures surrounding cytogenic testing. I am a professor with a doctorate in political theory, and my spouse has a BSc in occupational therapy, works as a senior health administrator, and earned a master's degree in theory and policy studies. We had ample qualifications to review both the technical and the ethical details surrounding the genetic testing that we were offered. Moreover, because we had two disabilities in the family – one serious, the other less so – we had significant exposure to the world of disability and concerns about the genomic scrutiny of fetuses. However, I discovered during the process that there were considerable contextual impediments to moral decision making in the real world. In the end, despite lengthy and intricate deliberations, our decisions were ultimately (and self-consciously) irrational and inconsistent.

I was adamant that, during our pregnancy in 2002, we would use amniocentesis.[2] I did not want our family to bear the financial, emotional, or physical burden of supporting yet another member's disability. It seemed prudent. However, by the day of the procedure, I was much less certain about this decision. Counselling associated with prenatal testing[3] had been fraught with difficulties. For example, during our sessions, no information about the various disabilities and conditions being tested for had been provided, and the counsellor had admitted that she made up statistics regarding the clinic's amniocentesis accuracy.

My spouse and I held long and arduous philosophical debates. Originally, she was more ambivalent about the procedure than I was because she was less certain that aborting a fetus in the aftermath of a poor result was an option that she could easily choose. The timing of amniocentesis (i.e., between fifteen and eighteen weeks of gestation), along with the two-week period waiting for results, means that women are potentially faced with undergoing a late-second-trimester abortion. Even I understood that this was a daunting possibility. Finally, we came to two important realizations: "We understood that there is a mutability to believing that a trait is part of natural diversity or, instead, is abnormal and therefore should be discouraged" (Atkins 2008, 112); and, in weighing our options, we were projecting ourselves into the future, we were imagining what we would and would not like or could and could not cope with, and we were in no way being altruistic (i.e., saving another's suffering). We were thus participating in an illusory cult of perfectibility.

In other words, a wide array of traits associated with gender, race, sexuality, and/or variant mental or physical traits can be seen as undesirable by parents and clinicians. As a family whose members have experienced discrimination based on gender, race, sexual preference, religion, and disability, we became leery of any clear demarcation of the normal from the abnormal – and we appreciated that this type of classification is essential to cytogenic testing. Furthermore, when we considered the lives of people with disabilities whom we knew, we fully grasped that neither were they miserable nor were their lives unworthy of living – for us, then, choosing prenatal testing did not mean that we would be sparing any future person an unlivable life. We were, instead, trying to spare ourselves imagined and anticipated suffering. We were being fundamentally self-interested. And, once we realized this, any lingering sense of altruism quickly dissipated.

Moreover, in trying to sculpt our future by using amniocentesis, we discerned that we would be taking part in a futile exercise of attempting to eradicate disability and illness from the life of our child (and, by extension, from our own lives). Even as we debated the ethics of screening for an imperfect child, we grasped the fact that birth itself carries risks and that life is fundamentally morbid. (For example, my brother-in-law is disabled because a car struck him when he was four years old. And, ironically, our current son – for whom we were considering cytogenic testing – has recently been diagnosed with a chronic disease.) Finally, I worried that our participation in genomic technologies would bolster the notion that we could (and should) engineer a more perfect human and thus contribute to the chronic revulsion and prejudice toward disabled (i.e., imperfect) human beings. Morally, neither of us wanted to condone practices that further alienated individuals with disabilities from mainstream society.

So what did we ultimately decide?

We used cytogenic testing for one pregnancy and rejected it for the other. However, we did submit to nuchal translucency screening[4] for the second pregnancy even though we thought such testing to be morally untenable. With regard to the first pregnancy, during the day of our scheduled procedure, all of the preoperative discussion about birth defects convinced my spouse that we needed to go ahead with the cytogenic testing given her "advanced maternal age." But by the time we arrived at our next pregnancy and our next decision, our moral convictions were stronger, and we presumed that we would resist the impulse to test our fetus.[5] These convictions, however, were eroded by the fact that the pregnancy was originally for triplets (and then later for twins), so our health-care providers scrutinized the gestation much more intensely. Although we ultimately refused amniocentesis, we could not turn away from nuchal translucency because it was a non-invasive look at the fetus through ultrasound and consisted merely of making a single measurement. It seemed irrational and counterproductive to reject this test since we had already undergone numerous ultrasounds. It also seemed counterintuitive to reject knowledge that was readily available, non-intrusively obtained, and seemingly risk free.

To be blunt, we were highly inconsistent in assessing and utilizing prenatal screening, and we made decisions that countered our own logic. But, paradoxically, if confronted by the same circumstances again, I believe that I would likely make the same choices.

In my previous article, I argue that contextual factors play a large role in the malleability of our thinking and decision making. Although I still believe that personal context did play a role in our uneven and ambiguous choices, I also believe that there is a symbiotic relationship between technological and genomic innovation and liberal doctrine in Western society. Liberal tenets of individualism, autonomy, and liberty do not act as moral curbs on the adoption of new technological and genomic products; instead, they facilitate and justify their uptake by individuals (and society). And, to make this connection clear, I need to revisit the origin of some of the central principles of liberalism.

As one of the originators of liberalism, John Locke is perhaps also one of its more articulate proponents. His *Second Treatise of Government*) clearly outlines a society in which men interact with the environment and "improve" it beyond the fallowness of its natural state: "God gave the world to men in common; but since he gave it [to] them for their benefit, and the greatest conveniences of life they were capable to draw from it, it cannot be supposed he meant it should always remain common and uncultivated. He gave it to the use of the industrious and rational" (Locke [1690] 1980, para. 34). For Locke, progress is an inevitable aspect of liberal civilization. Lockean men are equals and possess the freedom to labour and produce private property that improves their own lives and contributes to the "commodious" lifestyle of themselves and their fellows. Everyone benefits when a man encloses a plot of land and creates an orchard, for that orchard can now feed a community of men rather than remaining unused and uncultivated. In sum, Locke envisions an abundant earth and unlimited human industry. When these two aspects are united in an environment of moral equality and political liberty (i.e., in a liberal regime), they create a surfeit of wealth and choice, which in turn adds to the abundance of available resources and human productivity. It thus forms a metaphorical ascending spiral of "progress."

Yet, one can argue that one of the seminal liberals, Jean Jacques Rousseau, was deeply critical of the type of progressive, civilized society that Locke admired. In 1750, Rousseau maligned the manner in which "arts and science" corrupted men and allowed them to be morally flaccid as they became more sophisticated. Luxuries and leisure softened men's natural proclivity toward goodness and freedom (see *Discourse on the Arts and* Sciences). As a result, he opens *The Social Contract* (1762) with the highly evocative sentence "man is born free; but everywhere he is in chains." However, Rousseau argues in the same text that *reason* and *free will* distinguish man from the rest of the animals and give him a mastery over nature even as they can civilize him and render him miserable. In his vision of the best liberal society, our equality and rationality, along with our capacity to choose, are key. Self-governance is integral to liberal thought. The individual rules himself as he joins with other self-ruling individuals to form a collectivity of wills that, in turn, creates laws that govern society. The liberal man obeys laws that he himself forms. Political and moral authority thus resides in the individual (not in social conventions or divine authority). Although Rousseau's individuals form a "general will" in which all are united and free, his conception relies on the notion of the autonomous self. As Rousseau ([1762] 2006, 71) writes in *Emile*, "there are two sorts of dependence: dependence on things, which is from nature; dependence on men, which is from society. Dependence on things, since it has no morality, is in no way detrimental to freedom and engenders no vices. Dependence on men, since it is without order, engenders all the vices, and by it, master and slave are mutually corrupted." An ethos of rugged independence thus permeates Franco (as well as Anglo) liberalism.

Historically, Immanuel Kant furthered liberalism's egalitarianism by making universal reason the basis for man's equality. He believes that any man (educated or not) possesses reason and thus can access universal truths from which universal moral laws can be derived. His egalitarianism thus roots itself in the inherent rationality of all men. Furthermore, in his *Foundations of the Metaphysics of Morals*, Kant ([1785] 1996) differentiates between what can be known via phenomena and what can be known via noumena (i.e., what we understand through experience and what we understand through pure reason). In effect, he tries to simplify moral deliberation by divorcing it from contextual factors. Anything that is morally true must be found to be universally true (i.e., applicable to any situation). Kant creates a highly abstracted scheme in which the inner motive of an individual determines the moral character of what occurs. The willing of good is what matters, not its actual fulfillment. Something is deemed morally good because of the "goodwill" that incited it. Taken to its logical extreme, even if a man dies as a result of someone's actions, the murderer remains morally pure as long as what he willed was morally sound, even if the consequences were not. Kant thus values what is reasoned over what is performed or experienced. Consequently, Kantian liberal thought frames nature as a poor foundation for our moral deliberations and development. In essence, we raise ourselves up intellectually against the phenomena of nature and resist its and our own "natural" impulses. For Kant, reason allows us to construct a more perfect universe than the one that nature has provided.

John Stuart Mill develops the relationship between an individual's liberty and reason and his role in "civilization." In perhaps his most famous essay, *On Liberty*, he expounds on the importance (and social utility) of promoting absolute freedom of speech and conscience (opinion). Moreover, Mill clarifies the nature of a person's freedom:

The liberty of the individual must be thus far limited; he must not make a nuisance of himself to other people. But if he refrains from molesting others in what concerns them, and merely acts according to his own inclination and judgement in things which concern himself, the same reasons which show that opinion should be free, prove also that he should be allowed, without molestation, to carry his opinions into practice at his own cost. ([1869] 1991, 62)

The state (and others) must not interfere with an individual in those matters that concern the individual solely. A member of society can do as he pleases as long as it does not interfere with others.[6] And though Mill recognizes that some apparently self-regarding actions can have impacts on others and thus should be restricted,[7] his work propagates an image of a highly rational, bounded, and atomized self who acts alone and for itself, which is consistent with the work of earlier liberals.

Today, moral theorists and biomedical ethicists work within the liberal paradigm, and their writings echo the liberal philosophizing that has preceded them. Biomedical ethics and research ethics discourses present a familiar portrait of the individual and his or her choices in which theories of "deontology" reign. Here individuals have rights, and they determine their own paths. In medicine, a patient has the right not to have his or her body, industry, or attributes used for another's purpose without consent. The image of a rational, free, autonomous self who is accorded respect predominates the rhetoric. And, in this context, respect consists of allowing that individual a range of options and promoting the capacity to choose for himself or herself. For example, Peter Singer – a well-known philosopher – argues that it is morally sound for parents to terminate the pregnancy of a disabled fetus (presumably after genetic or ultrasound screening) and replace it with a more able zygote.[8] Singer (1993, 187-88) constructs a theory of "replaceability" in which the parents of a disabled fetus have the moral right to replace it with a more normal one that will have a better quality of life. For him, the desire to secure the most optimal future possible for one's offspring is natural and morally acceptable (as long as prenatal testing confines itself to the realm of disability and does not enter the realm of race, gender, eye colour, and so on).[9] Singer seems to remain unconcerned by the enormous range of traits that might be considered "disabling." Surely, in many societies, it is a disadvantage (even a disability) to be female, part of a minority ethnicity or race, or have same-sex orientation; thus, despite

Singer's excluding these as possibilities for "optimizing" one's zygote, it seems likely that others will not feel as squeamish about classifying such characteristics as "suboptimal" and thus worthy of exclusion for a future child. Consequently, the notion of individual right is employed to minimize or overlook some of the moral queries surrounding cytogenic testing. Of course, many oppose the genomic investigation of fetuses and gametes[10] because they find abortion untenable. They equate the life of the mother with that of the "unborn child," and thus an irresolvable contest occurs between one individual's rights and another's. However, even if one takes a pro-choice stance regarding abortion, genomic advances in detecting deviations from the norm in fetuses still raise troubling concerns. From this limited example, I hope that it is apparent that genomic technologies can facilitate the expression of a parent's right to choose the most optimal life for a child in his or her family. Scientific innovation thereby encourages, and is encouraged by, a liberal doctrine of individual right.

My review of key founding tenets of liberalism reveals that a number of them have a symbiotic relationship, rather than a critical one, with science and technology. Locke's notion of an abundant earth improved by the industriousness of rational and free men parallels a quest for greater productivity and innovation. Rousseau focuses on the manner in which a man can govern himself through his reason and will – it is this capacity that differentiates him from the other species on the planet. Although he creates the notion of the "general will" (i.e., the will of the people), it is composed of a collection of individual wills, which are independent and autonomous agents. He has contempt for those who are not self-sufficient. Kant further develops these ideas such that rationality (not nature, as in Locke, Rousseau, and others) becomes the basis for equality among men. Egalitarianism roots itself in a universal human ability to reason. Moral agency lies not in experience but in abstract thought. It is through the exercise of their minds that men improve impoverished nature – and science and technology are profound incarnations of this principle. Thus, both Rousseau's self-regulating individual and Kant's philosophizing moral agent correspond to an ethos of rational progress that inspires much scientific effort. Finally, Mill's proposal of a liberal self who has absolute liberty with regard to what concerns solely himself underscores the notion of an autonomous, independent person who advances society through the free expression of his opinion and the free ability to act as he pleases. Metaphorically, this type of individual shares much with an inquiring scientist or entrepreneurial innovator.

French philosopher and social critic Michel Foucault pointed to the connection between liberalism (and its rhetoric of equality, liberty, and brotherhood) and scientific positivism in many of his writings. He worked much as an archaeologist does, but instead of sifting through physical ruins he sorted through the archival materials of revolutionary and postrevolutionary France. By examining the records of prisons, hospitals, and asylums, he argued that the "freeing" rhetoric of liberal France masked a deeper, hidden vein of discipline. For example, the madman achieves "freedom" when he can master his behaviour so that he, and it, adhere to social norms. Thus, he experiences liberty only when he "disciplines" himself. In writing specifically about prisons, Foucault states that "a corpus of knowledge, techniques, 'scientific' discourses is formed and becomes entangled with the practice of the power to punish" (1995, 23).

A similar ontological entanglement occurred as my spouse and I weighed our options concerning prenatal testing. In particular, when we deliberated about the second pregnancy, we shared a belief that the promises of "able-bodied" children were illusory and denigrated persons who lived good lives with disabilities. Yet, when offered nuchal translucency, we agreed to learn about the statistical possibilities of Down's syndrome from the simple measurement. It seemed (and still seems) counterintuitive for us to turn away from knowledge gleaned from a non-invasive and non-risky procedure. In sum, we behaved like good liberals even though our previous moralizing went against our decision. Having knowledge, rather than not having it, seemed like the most responsible and best option. Furthermore, we gained our information at no one else's apparent expense. We severed ourselves psychologically from our earlier understanding that these types of decisions do have impacts on others – learning that our baby had Down's syndrome might have led to an abortion; even if it did not, the fact that we sought to classify our nascent child as falling within the normal range rather than the abnormal range denigrated those who are viewed as "abnormal."

Over the years, I have puzzled about my inconsistent ethical decision (and asserted elsewhere that context did impact our behaviour). I have also come to realize that my comfort with liberal philosophy actually encouraged me to make the choices that I did. The notion that I was an independent, rational, autonomous actor meant that I privileged these possibilities in my fetus. My impulse was to try to ensure that my future children embodied these values as much as possible. Furthermore, even when I sought to counter these urges, I failed to do so. In critiquing Kant, Foucault (1984, 37) writes that

Enlightenment, as we see, must not be conceived simply as general process affecting all humanity; it must not be conceived only as an obligation prescribed to individuals; it now appears as a political problem. The question, in any event, is that of knowing how the use of reason can take the public form that it requires, how the audacity to know can be exercised in broad daylight, while individuals are obeying as scrupulously as possible. And Kant ... proposes ... the contract of rational despotism with free reason: the public and free use of autonomous reason will be the best guarantee of obedience.

We obeyed our rational (liberal) impulses to know and have mastery over our lives (i.e., nature).

What I have concluded from this is that, as technological and genomic innovation becomes easier to incorporate into our lives, the more easily we will uptake these practices without critical reflection. The less invasive the procedure, the more easily it will align with our need to know and our drive to experience liberty as control over our lives and environment. When ethicists invoke the phrase "respect for the individual," they hope to make us pause and consider the dignity of that human being. The problem is that dignity is understood rhetorically to describe a person who can make autonomous, rational, free choices and act without constraints. This moral vision aligns easily with a scientific vision of the rational pursuit of objective facts and the fulfillment of these truths in the improvement of (our) nature.

Thus, even as I partake in research and clinical ethics committees, I am uncomfortably aware that the language and concepts that we employ are not as incisive as we imagine them to be. The liberal ideology and discourse that revolutionized human politics almost 300 years ago also provided the basis for the evolution of scientific exploration. For example, in *The Birth of the Clinic: An Archaeology of Medical Perception,* Foucault (1975) writes that postmortems were no longer forbidden in postrevolutionary (i.e., liberal) France, allowing physicians to finally understand the processes of disease as they saw the fruition of symptoms in the dissected corpse. In this instance, changes in the political arena thus dissolved moral prohibitions against medical-scientific investigation and discovery.

In sum, as we are increasingly expected to have opinions and partake in decision making about the use of genomic and technological innovations, we should become more acutely aware of the limits of our liberal values in the moral assessment of technology's emergence.

NOTES

1 I am currently writing a book entitled *Bad Moral Decision-Making: The Ethics of Everyday Life*. It outlines a host of seemingly mundane issues that beget deep and irresolvable moral quandaries.

2 We rejected chorionic villi sampling (CVS), in which a piece of the outer aspect of the placenta is biopsied, because it is associated with digit abnormalities in infants. For us, it was untenable to pursue a test for disabilities that had a correlation with inducing disabilities itself.

3 We live in Calgary, Alberta, Canada, so our care was funded through and provided by the provincial, public health-care system.

4 It consists of an ultrasound technician or physician measuring the fluid space at the back of the fetus's neck – if it is larger than the statistical norm, then the likelihood of Down's syndrome is higher.

5 Our opting out of prenatal screening is part of a larger trend. In a 2004 article examining prenatal care in Connecticut, the authors state that the use of these technologies fell between 1991 and 2002 despite an overall increase in maternal age (Benn et al. 2004). In another study, between 20 percent and 60 percent of pregnant women who received a positive blood-screening test for Down's syndrome declined amniocentesis and/or chorionic villi sampling to confirm or negate their serum results (Simms 2004).

6 In the twentieth century, Isaiah Berlin (1958) classified this arrangement as a form of "negative" liberty in which freedom is defined by a sense of non-interference from outside – it is about how the individual positions himself vis-à-vis others and the state and experiences an absence of intrusions or obstacles – whereas "positive" freedom, for Berlin, is about realizing one's inner impulses. Berlin argues that liberalism uses these two types of liberty interchangeably when it uses the rhetoric of freedom, but they can be in opposition with one another.

7 Mill ([1869] 1991, 90) writes: "If, for example, a man, through intemperance or extravagance, becomes unable to pay his debts or having undertaken the moral responsibility of a family, becomes from the same cause incapable of supporting or educating them, he is deservedly reprobated and might be justly punished; but it is for the breach of duty to his family or creditors, not for the extravagance."

8 Portions of this paragraph are drawn from a section in my article, "The Choice of Two Mothers" (Atkins 2008, 108-9).

9 Within this framework, Singer believes that fetuses are non-self-conscious beings that have neither reason nor a sense of their own future and thus do not suffer in the manner of more developed humans or more self-conscious animals, so their demise is ethically permissible.

10 The use of gametes here refers to pre-implantation genetic diagnosis (PGD), in which parents use in vitro fertilization and have their resulting gametes' cells examined for genetic abnormalities – only healthy gametes are then implanted into the mother's womb. This procedure is often employed by parents who already have a child with a genetically based disease (e.g., cystic fibrosis) to avoid giving birth to an affected sibling. Parents have also used this technology to find a donor match so

that a gamete's future umbilical cord blood can be used to harvest stem cells to provide a bone marrow transplant for an ailing family member.

REFERENCES

Atkins, C.G.K. 2008. "The Choice of Two Mothers: Disability, Gender, Sexuality, and Prenatal Testing." *Cultural Studies – Critical Methodologies* 8, 1: 106-29.

Benn, P.A., J.F.X. Egan, M. Fang, and R. Smith-Bindman. 2004. "Changes in the Utilization of Prenatal Diagnosis." *Obstetrics and Gynecology* 103: 1255-60.

Berlin, Isaiah. 1958. *Two Concepts of Liberty.* Oxford: Clarendon Press, 1958.

Foucault, M. 1975. *The Birth of the Clinic: An Archaeology of Medical Perception.* Translated by A. Sheridan. New York: Vintage Books.

–. 1984. *The Foucault Reader.* Translated by P. Rabinow. New York: Pantheon Books.

–. 1995. *Discipline and Punish: The Birth of the Prison.* Translated by A. Sheridan. New York: Vintage Books, 1995.

Kant, I. [1785] 1996. *The Metaphysics of Morals.* Edited by M. Gregor. Cambridge, UK: Cambridge University Press.

Locke, J. [1690] 1980. *The Second Treatise of Government.* Edited by C.B. Macpherson. Indianapolis: Hackett Publishing.

Mill, J.S. [1869] 1991. *On Liberty and Other Essays.* Edited by J. Gray. Oxford: Oxford University Press.

Rousseau, J.J. 1981. *The Essential Rousseau.* Translated by L. Blair. New York: Meridian.

–. [1762] 2006. *Emile.* Translated by B. Foxley. Charleston: BiblioBazaar.

Simms, M. 2004. "Screening Programme for Down's Syndrome." *British Journal of Midwifery* 12, 7: 454-59.

Singer, P. 1993. *Practical Ethics.* Cambridge, UK: Cambridge University Press.

Contributors

N/A

Chloë G.K. Atkins is currently an associate professor in the Department of Communication and Culture and teaches in the Law and Society Program at the University of Calgary. She has held Social Sciences and Humanities Research Council of Canada, Fulbright, Clarke, and Killam Fellowships. Currently, she holds an Ethics Catalyst grant from the Canadian Institute for Health Research that examines medically unexplained physical symptoms. Cornell University Press published her autoethnography, *My Imaginary Illness: A Journey into Uncertainty and Prejudice in Medical Diagnosis,* in 2010. The *American Journal of Nursing* awarded it the 2011 Book of the Year Award. She was also named the May Cohen Chair in Gender and Health in 2011 and delivered the Sue MacRae Annual Lecture for 2012. Currently, she is writing a book entitled *Bad Moral Decision-Making: The Ethics of Everyday Life.* She is also drafting another book manuscript entitled *Vulnerability, Trust, and Human Rights.* She researches and writes about clinical education, disability, sexuality, gender, medical ethics, and political and legal theory. She also serves on hospital and community clinical and bioethics committees.

David Castle is a professor and the chair of Innovation in the Life Sciences at the ESRC Innogen Centre, University of Edinburgh. He is a co-author of *Science and the Supermarket: The Opportunities and Challenges of Nutritional Genomics,* a co-editor of *Nutrition and Genomics: Issues of Ethics,*

Law Regulation, and Communication, and the editor of *The Role of Intellectual Property Rights in Biotechnology Innovation.*

Michael D. Cobb is an associate professor of political science at North Carolina State University. His research on public opinion and political behaviour has been published in the *American Journal of Political Science,* the *Journal of Politics, Science Communication,* and the *Journal of Nanoparticles Research.* He studies public opinion and citizen deliberation about science and technology, the intractability of misinformation about politics, and how the framing of casualties affects public support for war.

Keith Culver is a professor and the director of the Okanagan Sustainability Institute, University of British Columbia. Prior to joining UBC in 2011, he held the Econoving International Chair in Generating Eco-Innovation in the UniverSud Paris consortium of universities in Paris, where he developed graduate programs in sustainability and together with his team and industrial partners conducted research focused on the environmental performance of cities. Recent and forthcoming representative publications include Keith Culver and Kieran O'Doherty, eds., *Fishing and Farming Marine Iconic Species*; Keith Culver, Rachel Guilloteau, and Christelle Hue, "Hard Nodes in Soft Surroundings: A 'Dream of Islands' Strategy for Urban Sustainability" in *Development* 54, 3 (2011): 336-42; Keith Culver, *Legality's Borders* (2010); and Keith Culver and David Castle, eds., *Aquaculture, Innovation, and Social Transformation* (2008).

Peter Danielson is the Mary and W. Maurice Young Professor of Applied Ethics at the W. Maurice Young Centre for Applied Ethics, School of Population and Public Health and a member of the Cognitive Systems Program at the University of British Columbia. His research uses the methods of evolutionary game theory, agent-based modeling, experimental ethics, and public participation to contribute to a cognitive systems approach to ethics. He is the author of *Artificial Morality* (1992) and the editor of *Modeling Rationality, Morality, and Evolution* (1998). Information about the Current N-Reasons project is at http://yourviews.ubc.ca/.

Susan Dodds is dean of the Faculty of Arts and professor of philosophy at the University of Tasmania. She has written extensively in bioethics with a particular focus on public policy and feminist bioethics. She recently completed an Australian Research Council (ARC) Discovery Grant, "Big Picture

Bioethics: Policy Making and Liberal Democracy," with Rachel Ankeny, Françoise Baylis, and Jocelyn Downie and is currently working on an ARC Discovery Grant, "Vulnerability, Autonomy, and Justice," with Catriona Mackenzie and Wendy Rogers. Since 2005, she has been the Ethics Program leader within the Australian Centre of Excellence for Electromaterials Science, working on the social and ethical implications of emerging nano-technologies with a specific focus on bionic devices and nanomedicine. In 2010, she was appointed a member (ethicist) of the National Enabling Technologies Strategy Stakeholder Advisory Council.

Edna Einsiedel is a university professor and a professor of communication studies in the Department of Communication and Culture at the University of Calgary. Her research interests are in the social issues surrounding genom-ics and biotechnology applications and stem cell research. She has focused on approaches to public engagement and participation in these technologies and public representations of science and technology. She is currently a co-investigator on a GE³LS project on the VALGEN project (Value Addition through Genomics and GE³LS) and the PhytoMetaSyn project (synthetic biosystems for the production of high-value plant metabolites), both sup-ported by Genome Canada. She served as editor of the international journal *Public Understanding of Science* and was a member of the Board of Gov-ernors for the Council of Canadian Academies of Science until 2010.

John Gastil is a professor and the head of the Department of Communica-tion Arts and Sciences at Pennsylvania State University, where he special-izes in political deliberation and group decision making. His books include *By Popular Demand: Revitalizing Representative Democracy through Delib-erative Elections* (2000); *Political Communication and Deliberation* (2008); *The Group in Society* (2010); *The Jury and Democracy: How Jury Delibera-tion Promotes Civic Engagement and Political Participation* (2010); and the co-edited volumes *Democracy in Motion: Evaluating the Practice and Impact of Deliberative Civic Engagement* (2012); and *The Deliberative Dem-ocracy Handbook: Strategies for Effective Civic Engagement in the Twenty-First Century* (2005).

Colin Gavaghan is the New Zealand Law Foundation director of Emerging Technologies and an associate professor in the Faculty of Law at the University of Otago. He has published extensively on a range of subjects, including legal and ethical questions raised by reproductive, genetic, and nano technologies, end-of-life issues, and general medical-legal matters. He is the author of *Defending the Genetic Supermarket: Law and Ethics of Selecting the Next Generation* (2007).

Leonhard Hennen, Institut für Technikfolgenabschätzung und System-analyse at the Karlsruhe Institute of Technology, has a background in sociology and political science. He received his doctorate in sociology from the Technical University Aachen, Germany. After five years as a social researcher in the Department of Technology and Society at the National Research Centre Jülich, he was the project manager (1991-2005) at the Office of Technology Assessment at the German Parliament, run by the Institute of Technology Assessment, Karlsruhe Institute of Technology. Since 2006, he has been the coordinator of the European Technology Assessment Group (ETAG). His research interests include the sociology of technology, technology policy, concepts and methods of technology assessment, and TA in the field of biomedicine.

Michael Howlett is Burnaby Mountain Chair in the Department of Political Science at Simon Fraser University. He specializes in public policy analysis, political economy, and resource and environmental policy. He is the author of *Designing Public Policy* (2011) and *Canadian Public Policy* (2012). His articles and book chapters have been published in numerous professional journals and collections in Canada, the United States, Europe, Latin America, Asia, Australia, and New Zealand.

Michiel Korthals is a professor of applied philosophy at Wageningen University. His academic interests include bioethics and ethical problems concerning food production and environmental issues, deliberative theories, and American pragmatism. His main publications are *Philosophy of Development* (with van Haaften and Wren, 1996); *Tussen voeding en medicijn* (Between Food and Medicine, 2001); *Pragmatist Ethics for a Technological Culture* (with Keulartz et al., 2002); *Ethics for Life Sciences* (2005); and *Before Dinner: Philosophy and Ethics of Food* (2004). In 2010, he edited *Genomics, Obesity, and the Struggle over Responsibilities*. He has published approximately seventy peer-reviewed articles on philosophy.

Andrea Riccardo Migone teaches in the Department of Political Science at Simon Fraser University, where he has held a postdoctoral fellowship working on GE³LS issues in genomics research sponsored by Genome Canada and GenomeBC. He specializes in public policy, political economy, and globalization. He has published articles and book chapters on these topics in professional journals and collections in North America, Europe, and Asia.

Kieran C. O'Doherty is an assistant professor in the Department of Psychology, University of Guelph, where he convenes the Discourse, Science, and Publics research group. He has published on diverse topics such as public deliberation, genetic risk, the communication of uncertainty, human tissue biobanks, salmon genomics, and social and ethical implications of human microbiome research. Other areas of research and interest include health psychology, theoretical psychology, and the development and application of qualitative research methods.

Peter W.B. Phillips, an international political economist, is a professor of public policy in the Johnson Shoyama Graduate School of Public Policy at the University of Saskatchewan. He undertakes research on governing transformative innovation, including regulation and policy, innovation systems, intellectual property, supply chain management, and trade policy. He is the co-lead and principal investigator of a $5.4 million Genome Canada project entitled Value Addition through Genomics and GE³LS (VALGEN), which runs from 2009 to 2013. His latest book, *Governing Transformative Technological Innovation: Who's in Charge?* was published in 2007.

Alexandra Plows is an honorary research fellow at the School of Social Sciences, Bangor University, Wales, and an independent consultant. Her long-standing experience and interest in public engagement with eco and bio controversies is informed by her previous incarnation as an eco-activist. Her research interests include theorizing public engagement by using social movement theory to address questions of public expertise and engagement posed by science and technology studies. Her recent book, *Debating Human Genetics: Contemporary Issues in Public Policy and Ethics,* was published in 2010.

Grace Reid has completed a PhD and postdoctoral fellowship in science communication. She currently teaches scientific and technical communications to engineering students at Mount Royal University in Calgary, Alberta. She is also an adjunct assistant professor at the Centre for Research in Youth, Science Teaching and Learning (CRYSTAL) in the Department of

Educational Policy Studies at the University of Alberta. Her research explores the media's role in promoting public participation in scientific policy development. Her most recent research has been published in the journal *Public Understanding of Science* and an edited book collection entitled *Climate Change and the Media.*

David M. Secko is an associate professor in the Department of Journalism at Concordia University (Montreal). He currently runs the Concordia Science Journalism Project (www.csjp.ca), and his research spans journalism, science, and ethical issues to clarify and experiment with the roles of the public, experts, and journalists in the democratic governance of biotechnology. His recent articles include a qualitative metasynthesis of the experiences of a science and health journalist (*Science Communication* 34, 2) and a narrative analysis of online commentary after science and health stories (*Journalism* 12, 7).

David Weisbrot is a professor emeritus of law, an honorary professorial fellow in medicine, and a professor of legal policy in the United States Studies Centre at the University of Sydney. He was the president of the Australian Law Reform Commission (ALRC) from 1999 to 2009, where he chaired the landmark inquiries into the protection of human genetic information (*Essentially Yours*), gene patenting and human health (*Genes and Ingenuity*), and privacy (*For Your Information*). He is a member of the Human Genetics Advisory Committee of the National Health and Medical Research Council (NHMRC), and in 2010 he received the NHMRC's biennial award for the Most Outstanding Contribution to Health and Medical Research. He is a fellow of the Australian Academy of Law and was made a member of the Order of Australia in 2006.

Index

Note: GMO stands for genetically modified organism; "(t)" after a page number indicates a table, "(f)," a figure.

Printed and bound in Canada by Friesens

Set in Futura Condensed, Warnock, and Calibri
by Artegraphica Design Co. Ltd.

Copy editor: Dallas Harrison

Proofreader: Dianne Tiefensee

Indexer: Patricia Buchanan